Understanding Foucault:

A Primer for Beginners

With Appendices
on
Figuring and Indexing

by

David A. Harbour

Printed in the United States of America

ISBN 978-1-934916-01-8

June 2008 – Errata correction edition

NETZARI PRESS

Table of Contents

Foreword

Telescope makers are unique individuals who, gripped with the desire to make their own telescope will sometimes be smitten with a desire to grind and polish their own mirror. There is nothing like the sense of accomplishment of taking a raw glass mirror blank and grinding its surface into the superb mathematical curve which, when ultimately polished and then silvered (aluminized), will focus the light from distant stars, planets, and galaxies into tiny, perfect images to be examined by specially designed magnifying eyepieces! Finished mirrors are certainly readily available in the commercial marketplace, but *nothing* beats grinding and polishing one, yourself!

Mirror making literally requires hundreds of hours as determined telescope makers find themselves, among other things, carefully pushing a *mirror blank* over a matching glass tool slow grinding away the finest, microscopic glass particles to shape and curve the mirror surface into segment of a ***sphere***. After achieving the sphere, the surface then must be carefully reshaped, literally polished, into a ***paraboloid***. The paraboloid is critical to the performance of the mirror, indeed, to the entire telescope – without a near perfect paraboloid, the mirror simply cannot focus the light from a distant point source (i.e., parallel rays, focused to a tiny point – a very tiny, identical *image* of the source). Without that near perfect paraboloid, the image becomes a smudge, a blur, a less-than-perfect dot rendering the telescope useless.

When the spherical surface has been achieved, the next step is to polish the surface of the mirror with finer and finer grades of grit to slowly bring the surface into a fine polish. Only then, can the sphere be *changed* into a paraboloid by carefully changing the stroke of the polishing motion using the finest of polishing agents, typically rouge. To determine if you have achieved the paraboloid *you must apply some sort of test.*

Such a test, it turns out, was invented in 1858 by Leon Foucault, a physicist; the same man who, in 1851, used a pendulum to prove the Earth rotated. The assessment which now bears the name of its inventor, the *Foucault Test*, makes the shape of the curve of the mirror visible to the naked eye! Actually, it is not the shape that is visible; rather, it is the ***deviations*** from the desired shape which are rendered visible.

The Foucault test for parabolic mirrors is the "pinnacle" test for telescope makers to visually assess the results of their hard work. Did they reach the paraboloid or not? Did they achieve the paraboloid or did they create the dreaded "turned-down edge"? (A turned-down edge destroys the usefulness of the outermost part of the mirror, which is devastating as the outermost part of the mirror is responsible for gathering the most light; and with a turned-down edge, you may as well have built a smaller telescope at the start.)

Telescope makers, being the hardy stock they are, will also build their own Foucault Tester with which to conduct the test. It is not as hard to make this tester as it is to build many other parts of the telescope because it does not present an engineering challenge. Basically, a holder for a knife-edge, and a fine mechanism to move the knife-edge in very small but controllable increments is all that is required. Attaching a fine scale (a micrometer screw), permits you to record the precise location of the knife-edge along the optical axis. Even the stand to hold your mirror upright, and fixing a light bulb to the tester to simulate a distant star do not present difficult challenges.

The great difficulty comes in the *conduct* of the Foucault Test and *interpreting* the results! Here is where it pays to have a trained eye and training and education in optics and mechanics. Here is where you must possess skills to interpret shadows – the inky-grey shadows that must be assessed to determine if they are "in" or "out" or "high" or "low". Here, you need help.

Help has come.

David Anthony Harbour, my good friend of many, many years has written what must be the most succinct and lucid description of the Foucault Test that anyone may have ever written to date. He presents not the usual, bland, mechanical treatment of the subject, but rather a dynamic and detailed description of the technique such that the amateur telescope maker, the "ATM," both inexperienced and experienced, is able to conceptualize the appearance of the surface of the mirror and know exactly what to do if the mirror must be returned to the work stand! That, after all, is the purpose of the test - to reveal deviances from the true sphere and understand when the shadows represent the paraboloid! For what good is a test if, after conducting it, you have no concept of what the test means or, worse, what to do about it?

In the short narrative which Dave entitled "Understanding Foucault," he takes the amateur through a comprehensive understanding of the Foucault test and its merits. He introduces the test in a series of "essential foundation lessons" where the reader is provided a priori understanding required for interpretation of the Foucault test.

From this brilliant work, any amateur, after building his own Foucault testing device, can now understand what he or she sees in the wispy shadows of the mirror which is destined to become part their great telescope. The reader will *understand* what he is seeing such that he will be able to interpret the information and apply corrective actions necessary to achieve that perfect mirror! I am confident, that no other description of the Foucault Test can lead the telescope maker so aptly toward understanding the results of a Foucault test, and make it clear what to do about it.

Dave Harbour deserves great credit for this narrative. It is for this purpose that I decided to publish it. I wished to honor my good friend with his name permanently emblazoned as its creator and author on the cover of this book; ***his*** work – "Understanding Foucault." Dave has permitted this work to be published on the

World Wide Web and it has been formatted and prepared by several telescope making web site owners and is freely available by simply accessing these sites. But this work, deserved to be printed in traditional hard-copy.

I only wish that the amateur telescope making world could have experienced the amazing brilliance which I enjoyed while closely working with David Anthony Harbour during the years that he and I designed and built telescopes together, and in the countless hours we spent observing in the clear night sky of Northwestern Oklahoma in the 1970s. Dave has a phenomenally sharp mind, and in him lies the talent and vision of an artist with the inquisitiveness of the scientist. He possesses an incredible ability to grasp cause and effect relationships and the ability to recall and apply obscure facts in a broad range of technologies including, of course, telescope making; but also photography, mechanics and machinery. His expertise and in-depth knowledge includes aviation both as a pilot and as a student of the theory of flight. He has studied dirigibles and zeppelins and can recite the most fascinating facts about the era of lighter-than-air. The list could go on and on.

In honor of David, the *Leonardo's Discovery Warehouse & Adventure Quest* in Enid, OK, has prominently displayed a striking, giant stainless-steel sundial he designed and built which, by its uniquc design, can tell the time to the nearest minute either in standard time or daylight savings time. One can also find a solar heliostat built by David in the *David Harbour Solar Observatory* on the campus of Northern Oklahoma College.

In this book, I have taken the liberty of adding to "Understanding Foucault" three additional Dave Harbour narratives - instructional guides which he wrote to help the ATM complete the mirror. The first is "Figuring the Paraboloid" which provides a method for achieving the parabolic shape in the mirror surface. The second builds on figuring by the application of the "Area Formatted Ring Lap." And the third is his article on "Indexing" in which Dave describes his "idiot-proof" method of ensuring that the figuring of the mirror is accomplished in a uniform manner, resulting in the homogeneous application of the polishing actions required to achieve the desired figure on the mirror. Dave wrote these articles to assist the ATM and it was my desire that these articles be captured in this book along with his foundational work, *Understanding Foucault* and made available to ATMs around the world.

William J. Welker
Colorado Springs, Colorado
March 2008

Understanding Foucault:

A Primer for Beginners

Introduction

An experience of rare delight awaits one who beholds a celestial wonder with a mirror crafted with his or her hands. Most persons making a mirror also find another kind of rare flavor in the experience: the reward of meeting a challenge that requires discipline and concentration of one's mental faculties.

Almost certainly the last phase of mirror making, where one imparts the final, precisely defined curve to the mirror's reflective surface, is the most challenging. This final stage of mirror making is known as *figuring,* inasmuch as it is a process of altering the mirror's curve, or *figure* into the one required for proper function. The process of figuring is essentially a process of selectively removing very minute amounts of glass in the proper places to bring its curve to the required form. In order to guide the process of figuring we need a sensitive test that will accurately show us the mirror's true figure at every stage of the work.

With this treatise, I will show that *for the amateur* Foucault has altogether more advantages and fewer disadvantages than any other method of testing. The method is extremely versatile and more than adequately precise. Foucault is even adequate enough for the production of a fine primary mirror for the classical Cassegrain reflector. I have written this exposition in a new way in order to make the basic foundation concepts understandable in as intuitive a way as possible, not assuming any previous familiarity on the part of the reader with any of these founding concepts.

For this reason it seems appropriate to insert a necessary note for those experienced and knowledgeable persons who may want to offer comment on this exposition. You will find many illustrations, descriptions, and explanations that will strike you as perhaps incomplete, or not quite precisely literal, or not exhaustive of the subject. I ask all of you to bear in mind that I am trying to educate the beginner, rather than confuse him or her with too much unnecessary information. My goal is to teach first principles first.

Testing in General

Monitoring the developing figure of a mirror can be done with a wide variety of methods. All of these methods fall under two basic genera with their several different species: tests done at center of curvature, and tests done at focus. We consider testing at conjugate foci as a variation of testing at center of curvature.

Testing at focus is very convenient and is the preferred method for many advanced amateurs. The commercial houses also usually prefer it. For the amateur with limited resources wanting to make only one, or a few mirrors, it is not so suitable as testing at center of curvature. This is because both of the best forms of testing at focus, null testing and interferometry, require auxiliary optics that are either prohibitively expensive, or overly challenging for the beginner to fabricate. Thus, we will conclude that testing at focus is not convenient for the purposes of beginners.

That leaves us with that other genus of testing with its various species: tests done at center of curvature. The best forms of this type of testing fall (at least approximately, if not precisely) under two basic forms: null tests, and non-null, quantitative tests.

Null testing at center of curvature we dismiss as not convenient for the purposes of the beginner for the same reason we dismissed it as a testing method for use at focus: it requires auxiliary optics either expensive or difficult to fabricate. In addition, this form of testing introduces an extra level of opportunity for error- I will merely cite the tragic case of the Hubble telescope's seriously flawed optics for an example of null testing at center of curvature gone seriously awry, even in the hands of seasoned experts.

[Ed: It may now be long forgotten, but the tragedy of the Hubble Space Telescope refers to the misfigured 94.5 inch diameter primary mirror made by Perkin-Elmer in the early 1980's. The primary, because it would operate in space with no atmospheric interference, was supposed to be "the most perfect mirror ever made", instead was produced with severe spherical aberration. The flaw was not detected, at least not with certainty, until after the telescope was launched into orbit. The central cause of the flaw was determined to be due to a mis-positioning of the optical instrument used to guide the polishing. The instrument was off by only 1.3 millimeters! The mirror did, indeed, measure the mirror's figure as perfect – and it was! But since the instrument in the null test was ever so slightly misplaced, the perfect shape was not the ***required*** shape! The flaw would have been detected by the Foucault Test.]

At last we come to that other species of tests at center of curvature: non-null, quantitative tests. There are two best, most useful sub-species of this kind of testing: the Gaviola, or so-called "Caustic" test, and lastly, Foucault.

A well known, "great light" in the optical industry once dismissed, in conversation with me, both Caustic (and Foucault) as "useless". After finally mastering both these forms of testing, I understood why he'd made this remark: he used interferometry, a far less cognitively challenging method for testing, and did not understand the other methods.

Caustic is capable of monitoring the developing curve on a mirror being figured to an extremely high degree of accuracy. This very precise, rigorous form of testing is, however, both time consuming and tedious in its execution. Additionally, the test apparatus is expensive to purchase, or overly challenging for one without machine tools or machining skills to fabricate. So we will pass on Caustic as also not suitable for the beginner's purposes.

We finally come, then, to Foucault. In defense of its merits as weighed against its demerits, I refer the reader again to my general comments in the second paragraph of my introduction. Additionally, I would like to say in praise of Foucault that it is peculiarly "idiot proof" in a way that null testing at center of curvature is not. Again, I cite the near disaster of the Hubble telescope as an example of the kind of error that can happen with null testing at center of curvature. This kind of error can't happen with Foucault. Lastly, I want to add that neither machine tools nor machinist's skills are required to construct a very adequate test apparatus for Foucault. Time has tested Foucault - "Nothing tests truth like time".

Let's make a telescope mirror!

We Need a Mirror

In order to begin learning Foucault, we need a concave mirror. I cover the mechanics of grinding and polishing the preliminary curve onto the glass in another paper- we are interested here in learning testing so that we can figure the final, correct curve onto the glass to make it give sharp images.

We're going to make our mirror in an imaginary way. The reason we're going to do this is to teach you your very first *essential, foundation lesson* for understanding Foucault- to help speed you on your way to a complete mastery of the required concepts.

Perhaps some civilization beyond the earth, with an advanced technology, makes mirrors by a method we shall postulate. Maybe the vanished "Krell" civilization on Altair 4 made their mirrors in the way we shall describe. The Krell opticians will demonstrate how they will make the initial concave curve for a twelve inch mirror with a very short focal length. For our first lessons in our treatise on Foucault testing, a very short focal length mirror serves for clarity in the illustrations better than a longer focal length mirror. We will pretend that Foucault will serve well for testing such a very short focal length mirror- even though we know that Foucault is not satisfactory for testing mirrors of extremely short focal length. Remember that our purpose is to teach.

Molding a Mirror

Perhaps archeologists will discover that the long vanished Krell opticians did not grind their mirrors, but molded them. Let us

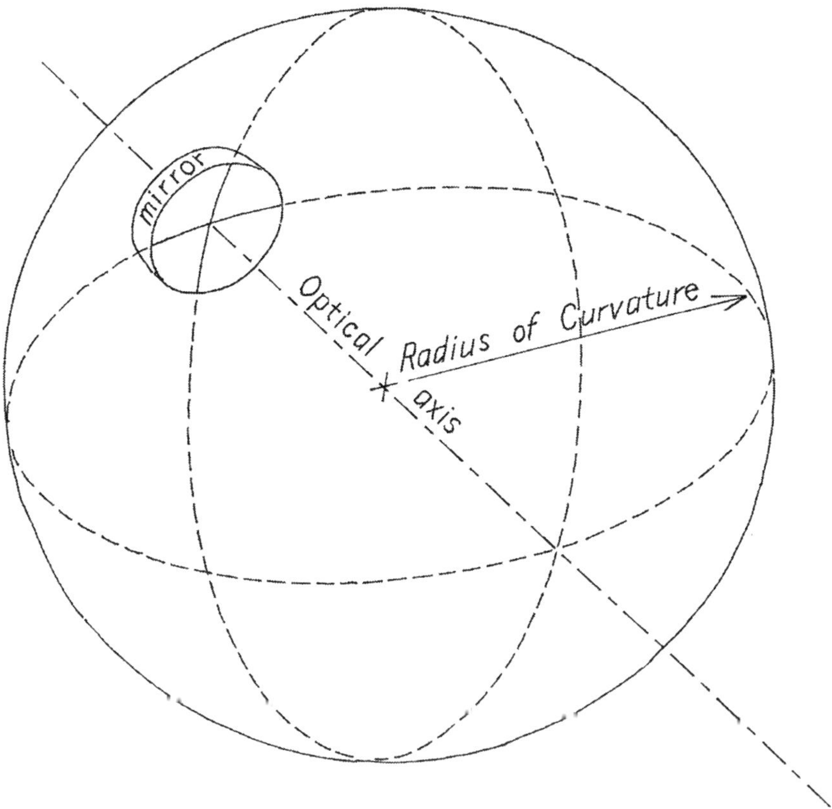

Figure 1.

postulate that they used a large, perfectly spherical ball of their fabulously durable "Krell metal" to impress the initial concave curve onto a mirror being made. We will refer to Figure 1 as we discuss this molding technique. Looking at Figure 1, we see a depiction of a large sphere, drawn as if it were transparent. A dotted line of longitude, and a similar line of latitude, are drawn onto the surface of the sphere to help convey its three dimensional shape. On the far side of this apparently transparent sphere of Krell metal we can see a telescope mirror pressed up against its smooth, spherical surface. And this is the first step in Krell mirror making: the initial curve is impressed onto the surface of the glass after it is heated to near melting by pressing the mold with the hot glass up against the surface of the big sphere. One can plainly see that the shape of the curve *impressed* onto the glass with this molding procedure is an *exact* representation of the shape of that section of the sphere's surface against which the mirror was pressed into contact.

1st Essential Foundation Lesson:

Now this method of molding a mirror helps us understand some things. Firstly, it helps make clear what is meant when we say that this mirror's curve is *spherical;* of course it is spherical - words are not really needed to describe its curve - its shape is a direct impression of the big sphere of Krell metal. So we have a spherical mirror now, which after further work will become a telescope mirror. Let's look at some features of this spherical mirror. We understand now that it has a *spherical* shape, or *figure,* after being molded with the big sphere of Krell metal. We will often have occasion to refer to its figure's *radius of curvature.* Please look at Figure 1 again; it is immediately apparent what we mean by *radius of curvature* when referring to this mirror's spherical figure; its radius of curvature is identical to the radius of curvature of the surface of the big ball of Krell metal used to mold it. So our imaginary molding technique for impressing the initial concave curve onto our mirror has taught us two things:

Essential Foundation Lesson #1 - ***What is meant by the term spherical, as a description of a mirror's figure, is that the shape is that of a section of an imaginary sphere and that when we speak of its radius of curvature we are referring to the radius of this imaginary sphere.***

2nd Essential Foundation Lesson:

While learning this next lesson, we will be referring to Figure 2a and Figure 2b. Figure 2a shows a view of our newly molded mirror with its concave, spherical figure in cross section, receiving four parallel rays of light as they enter from the right side of the diagram. These rays of light are parallel because they originate from a very small source of light a very great distance away from our mirror- we will specify that the light source is infinitely far away. The rays of light striking the areas of the mirror's surface near its edge are shown as dashed; the rays of light striking the mirror's surface near its central regions are shown as unbroken. The line shown proceeding through the center of the face of the mirror we shall call its *optical axis.* Strictly speaking, a spherical mirror does not have an optical axis, but we will pretend that it does, for teaching purposes.

Now, looking at Figure 2a, we can see that the light rays are reflected from the concave spherical surface of our mirror and *focused* onto a specific region along the mirror's optical axis. A casual glance suggests that the mirror will focus all four light rays down onto the same point lying on the optical axis, but a close inspection will reveal that this is not quite the case.

The rays of light that were reflected from the outside regions of the mirror's concave spherical surface (represented by the dashed lines) come to a focus at a point lying on its optical axis slightly closer to the mirror than those rays of light reflected from the interior regions

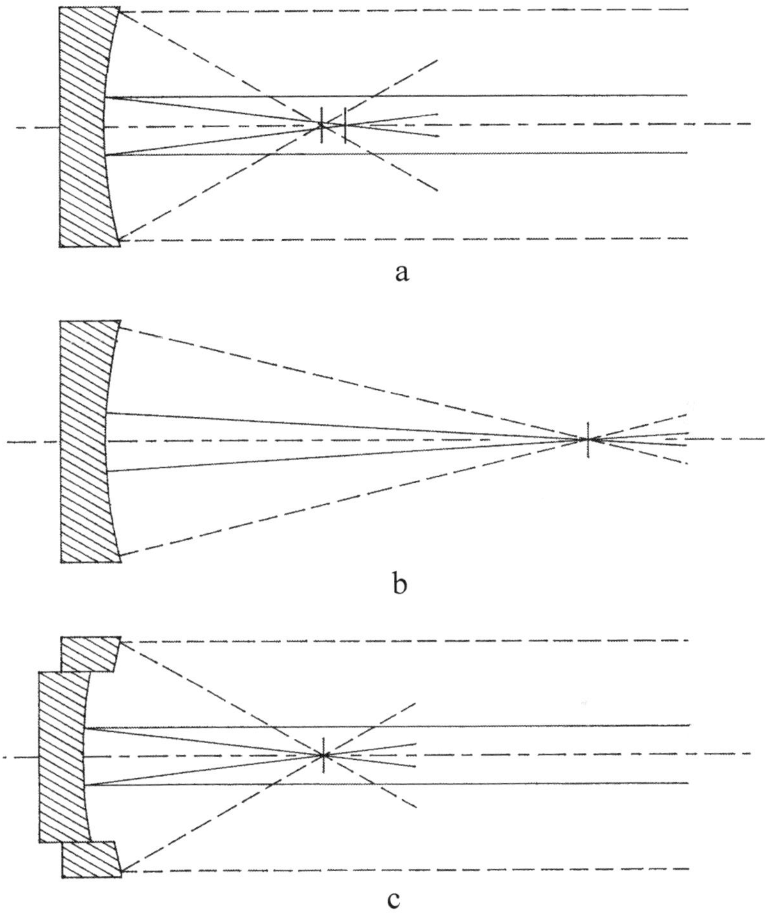

Figure 2 a, b, c.

of the mirror's concave spherical surface. In our illustration, two short, vertical lines mark these two different locations along the optical axis. We will think of these two short, vertical lines as representing two different locations of a small square of ground glass, on which to see the focused image of the infinitely distant light source. Or, just as usefully, we might think of them as representing two different locations of a small square of photographic film, with which to record the focused image of the infinitely distant light source. What is essential for us to understand here is that a mirror with a spherical figure (we will call this kind of mirror a "spherical mirror") receiving light from infinity cannot focus all of this light reflected from its inner and outer zones into the same focal plane. This characteristic is known as *spherical aberration* (naturally enough).

The mirror in our example as molded by the Krell in their optical plant has a very short focal length; really, its focal length is even too short for us to use it in a telescope. We have made it very short so that the diagrams will be clearer and more instructive. When we speak of a mirror's focal length, we mean, of course, the distance from the mirror's concave, reflective front surface to the point (or, more precisely, a field of all possible points, a "focal plane") into which any given bundle of parallel rays received by the mirror from infinity will be focused. In short, the focal length is the distance from the front surface of the mirror to its image, or focal plane. The diameter of a mirror, divided into its focal length, gives us its *focal ratio.* We will pretend that our Krell opticians have molded us a 12" mirror with only a 12" focal length. Thus, the focal length of our mirror is unity, and expressed (by convention) as f/1.

While we are still considering Figure 2a it will be helpful, here, to understand that this illustration can be thought of as representing only how a mirror makes an image of a very distant, apparently very small, *single light source,* such as a single star that the telescope is aimed directly at. A little thought will make it apparent, however, that for an image field comprised of more than a single star, such as for instance a galaxy, a mirror will be receiving many different bundles of parallel light from many different points of origin in its field of view, each of these bundles approaching the mirror at a slightly different angle, with each bundle reflected at a slightly different angle from the mirror and focused into separate, disparate points in the focal plane, building up an image all across this plane.

Now, let's look at Figure 2b. For this part of our lesson we have moved the infinitely distant single, small light source up very close to our mirror but still lying on its optical axis. We have brought it up to a point where light rays fanning out from this very minute light source will each strike the mirror at an angle that we will loosely describe as a right angle. After striking the mirror, each will be reflected back at the very same angle, returning exactly along its outgoing path back to its source. For a spherical mirror, there is only one point along its optical axis where the light source can be located for which the mirror will reflect (and focus) its spreading rays exactly back onto it. The location of this point is identical with the mirror's center of its radius of curvature. And now we've learned in detail:

> **Essential Foundation Lesson #2:** ***A spherical mirror, receiving a bundle or bundles of light rays from infinity, cannot focus all of the light rays in each bundle into the same image, or focal plane. However, if the light source is located at the center of its radius of curvature, it will precisely focus all the source's light rays back onto its center of curvature.***

Understanding this feature of the spherical mirror's optical properties will help us later on master necessary concepts for Foucault.

Next Lessons

We will not call this next instruction an "essential foundation lesson", but it will prepare you for one.

We've seen that a spherical mirror cannot focus light from infinity that has been reflected from different areas of its surface all into one focal plane. This defect is known as *spherical aberration* and will prevent the mirror from focusing light accurately enough to form a sharp image.

Fixing the Defect

How can we possibly remedy this defect, this inability of a spherical mirror to accurately focus all its reflected light into one focal plane? If we look again at Figure 2a we can find a clue as to one possible remedy- since the rays of light that are being focused by the near-central regions of the mirror are coming to a focus farther away from the mirror, why can't we just move the central portion of the mirror back a ways, until its focal plane is congruent (in the same location) as the focal plane for light focused from the outer (near mirror's edge) regions? Figure 2c shows how this remedy might work: note how the central portion of the mirror has been parted from an outer annulus, and moved backwards by a distance required to bring its focal plane congruent with that for the outer annulus. The remedy was actually tried by a famous nineteenth century telescope builder, Lord Rosse, whose chief contribution to astronomy was his giant reflector at Birr castle and his discoveries with it. The remedy is impractical for several reasons, however, and probably tried only once. But I have shown you Lord Rosse's experiment for instructional reasons- we are going to use his curious, once ever two component mirror to learn an interesting lesson that will help us understand Foucault. Before we meet our "Lord Rosse Special" again, however, we need to learn about the *knife edge.*

Leon Foucault

The nineteenth century saw many important foundation developments that contributed to the perfection of the modern reflecting telescope. Leon Foucault (1819-1868) contributed some of the most important innovations. In addition to demonstrating the superiority of glass (with silver coating) over speculum metal for mirrors, he presented the world with a powerful method for critically surveying optical surfaces- the knife-edge. In his hands and the hands of his followers various developments and applications of the use of the knife-edge as an adjunct for accurately figuring optical surfaces were developed over time.

The Knife Edge

Let's look now at Figure 3a. We have our 12" spherical mirror, molded for us by the Krell opticians, standing on edge. We've left the two meridians of longitude and latitude, impressed onto the mirror's front surface from the big ball of Krell metal, to remind us of the mirror's concave shape. Four rays of light are depicted as being focused by our mirror down to a point, apparently just adjacent the edge of a small razor blade standing on its end. Please note that the edge of the blade is extremely close to the mirror's optical axis. These light rays have emanated from a very tiny light source located on our mirror's optical axis at the center of its radius of curvature and fanned out to its concave surface, to be reflected back along their outgoing paths to converge right back onto their tiny light source. After converging onto their original source, they continue on, fanning out and away. Although we show only four light rays fanning out from this tiny bulb, in order to keep the diagram uncluttered, we will keep in mind that in reality there are light rays fanning out from this little light source in great numbers in every direction and are striking the mirror's front surface all over.

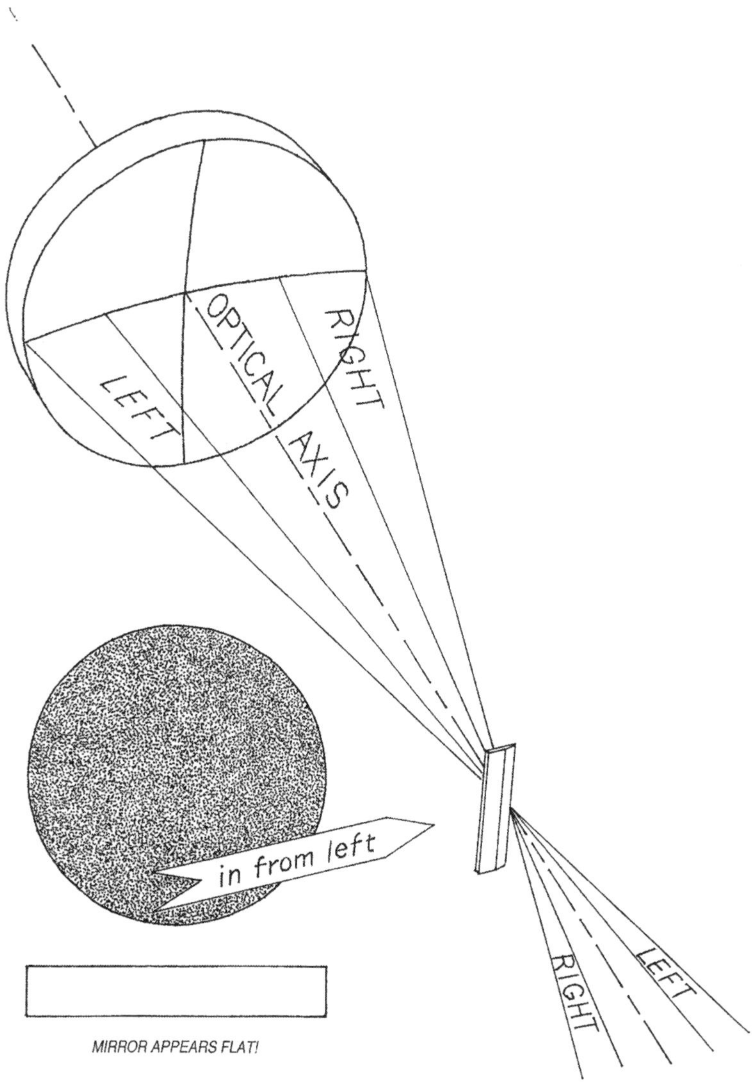

Figure 3a.

In this illustration our tiny light bulb at center of curvature and the razor blade's edge are virtually in the same location. In order to keep the illustrations as uncluttered as possible, we are going to adopt the convention of not showing the light source at center of curvature.

The mirror is actually forming an image of our tiny light source right at center of curvature, and even if we do not have a piece of ground glass or film in place here for the image to be focused on, it will still be in place here, just suspended in the air- a so-called "aerial image".

Now we are going to do an experiment. First, we will move the razor blade away from the optical axis to the left, slightly- just enough so that we are sure it is not obstructing any of the reflected rays of light crossing over the optical axis here at center of curvature (we will shorten "center of curvature" to "C of C" from here on out). Next, one of us places his or her eye in position on the mirror's optical axis just behind and very close to the tiny aerial image formed there at C of C. Now, with one's eye placed very close to this tiny image, one's eye will not be able to focus on it and see it, but rather, one will see the mirror beyond it. The mirror will appear brightly and evenly illuminated all over, because it is reflecting the light from the tiny light source equally from every portion of its surface.

3rd Essential Foundation Lesson:

Next, we begin to bring the razor blade in from the left, very slowly, until its edge begins to obstruct, or occult part of this light returning from the mirror. Now, inasmuch as *each and every part* of the tiny image of the light bulb is receiving a bundle of light originating from *the mirror's entire surface,* when the razor blade begins to obstruct it, the light in each of these bundles from the mirror is reduced equally and simultaneously from every part of the mirror's surface, and the mirror begins to darken all over simultaneously and equally, "graying out" all over, uniformly.

We may halt the advance of the knife-edge (KE) when it is obstructing about half of the returning light from the mirror; or, we may continue its advance until the returning light is entirely cut off. If our mirror is accurately spherical, and the KE approaches and crosses the optical axis at exactly the C of C, the mirror will darken, or *null,* simultaneously, all over its surface. When we use the KE to examine our spherical mirror from the vantage point of its C of C this way, the mirror will *appear* perfectly flat. We know for certain that it is spherically concave, but nevertheless it appears flat when viewed this way. And as it turns out, there are important advantages for us in pretending, or imagining the mirror as flat, instead of concave. And this is the essence of another:

> **Essential Foundation Lesson #3:** ***We will adopt the convention of always visualizing the mirror's surface as flat, and any deviations from this imaginary flat figure will always be visualized and depicted as such.***

4th Essential Foundation Lesson:

Now, let us consider Figure 3b. Our razor blade (KE) can be moved at right angles to the mirror's optical axis, and also parallel to its optical axis, toward or away from the mirror.

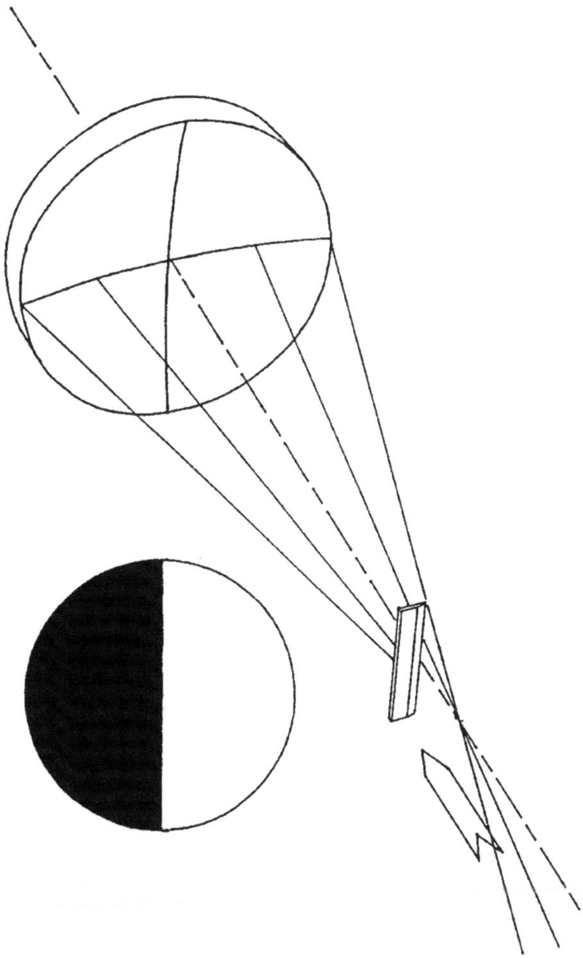

Figure 3b.

In Figure 3b we have moved it a little ways toward the mirror, leaving its edge just adjacent the optical axis. The KE is now blocking only those rays of light coming from the left side of the mirror, and only its left side appears dark. If we withdraw the KE to the left, away from the optical axis, this dark shadow on the left side of the mirror will recede to the left, also. If we advance the KE back in again to the optical axis, the shadow will reappear and advance in the same direction across the mirror's surface. If we continue advancing the KE all the way across the optical axis (OA) until all of the returning rays of light are blocked, the entire mirror will go dark as the KE's shadow advances all the way across the mirror from left to right.

Now, let's back the KE away from the mirror along its OA, passing through C of C until we are beyond it by about the same distance as we were just previously inside it (closer to the mirror than C of C) as shown in Figure 3c (next page). Suddenly, even though we have not moved the KE laterally, the right half of the mirror now appears darkened! Look at the illustration carefully- with the KE beyond C of C, it is now blocking the rays of light from the right half of the mirror after they've crossed the OA. The light rays from the left half of the mirror, however, have crossed over the OA in the other direction, away from the KE, and are not obstructed by it at all. Now, if we back the KE out to the left again, away from the optical axis, the shadow on the right side of the mirror will advance in the opposite direction of the KE, towards the right.

If we back the KE out all the way to the left, so that it no longer obstructs any of the returning rays from the right half of the mirror, the mirror will again appear bright all over. Alternatively, if we advance the KE back in again from the left until all of the rays of light returning from the mirror are obstructed (occulted) then the mirror will go dark all over as the KE's shadow advances in from the right, again moving in the opposite direction as the KE.

We've just made a wonderful discovery, and learned:

> **<u>Essential Foundation Lesson #4:</u>** ***The KE can tell us whether we are inside of, outside of, or exactly at the center of curvature of a spherical mirror.***

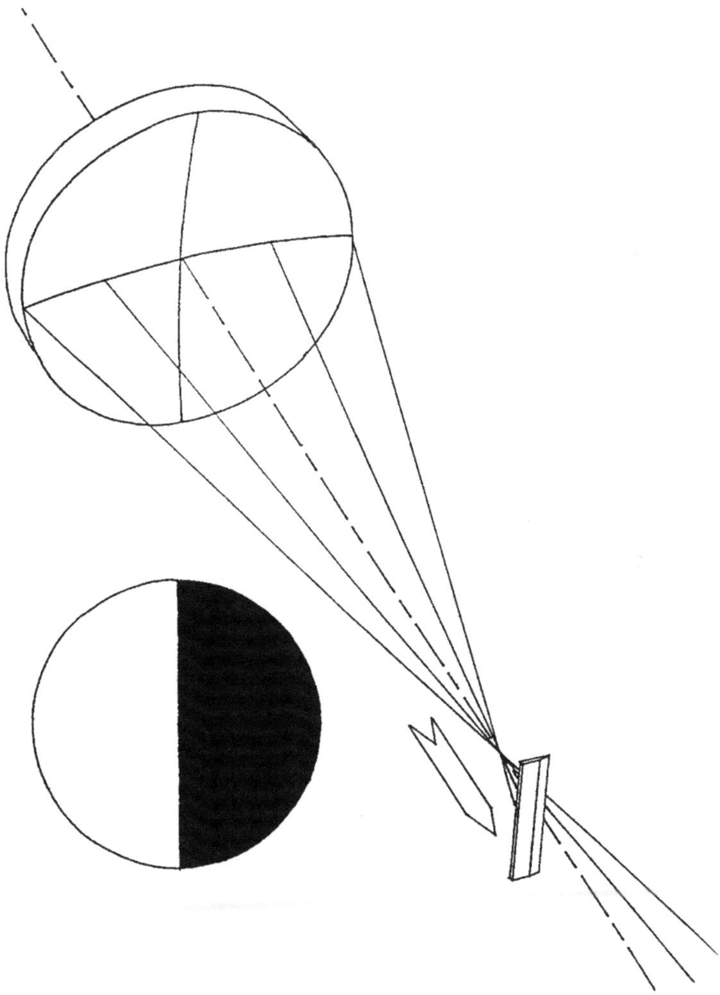

Figure 3c.

The "Lord Rosse Special"

We've learned that the knife-edge can tell us just about one important thing about a perfectly spherical mirror: where its center of curvature is. And although spherical concave mirrors have their places in astronomical optics, we are primarily going to concern ourselves with another species of concave curve that is not as simple as the sphere. Remember- a spherical mirror suffers from *spherical aberration*- a characteristic that prevents it from accurately focusing light from infinity, preventing it from forming sharp images.

Remember our discussion about Lord Rosse's efforts to cure spherical aberration with his curious two component mirror? We want to use his novel mirror to learn a new, essential foundation lesson.

Let's take a look at Figure 4 on the next page. It shows us Lord Rosse's two component mirror standing on edge. We've put two meridians across its front surface to help us visualize its shape. This mirror was made by parting a single, one piece spherical mirror into two components.

Before it was parted and the two components relocated slightly with respect to each other, it had, of course, a single radius of curvature and therefore a single center of curvature. But now this "compound" mirror has two different centers of curvature. They are marked in the diagram as two short lines lying across the optical axis in slightly different locations, at "A" and "B". Two unbroken lines representing two rays of light are shown emanating from the C of C marked "A" and fanning out, striking the interior of the central component of the mirror. After being reflected from this area, they converge back on their C of C, crossing the OA at that location and then fanning out beyond. Four dashed lines, rays of light, are similarly shown fanning out and striking the outside component or annulus of the mirror and then being reflected back to their C of C, crossing the optical axis there and fanning out beyond. Let's take a

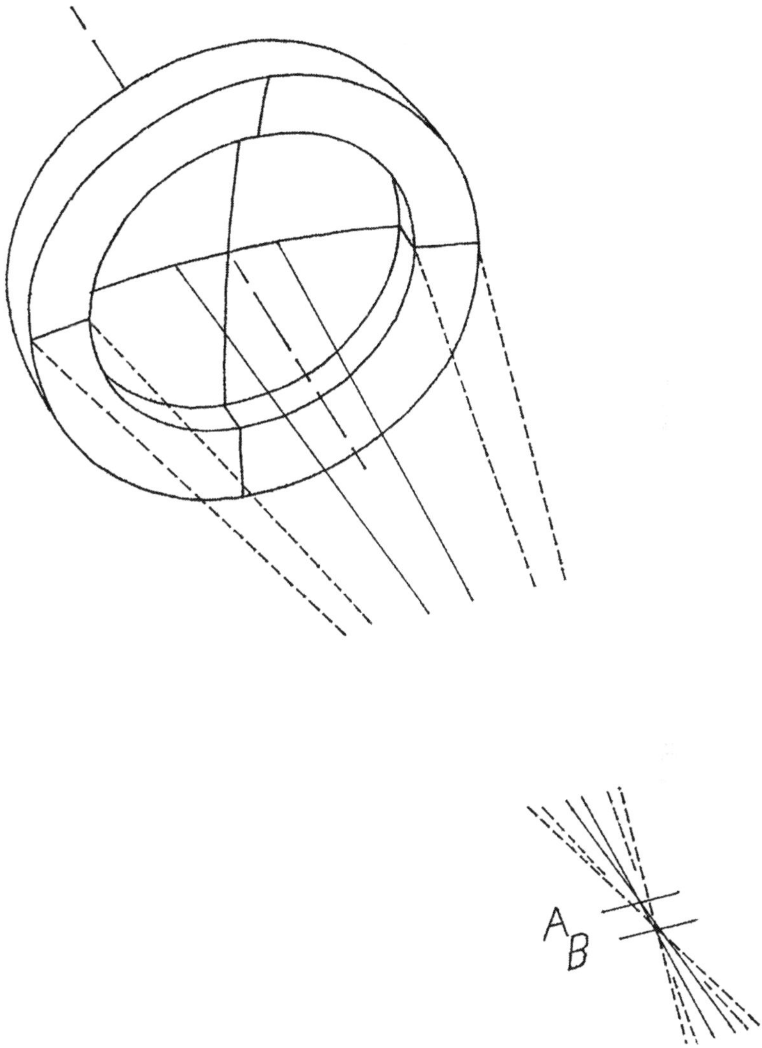

Figure 4.

close-up look at this region of the OA where all of these rays are crossing it.

In Figure 5a we've zoomed in close to see what's happening more clearly where these light rays are crossing the OA. At each center of curvature for each component of the mirror we've positioned a little square of ground glass (or we may think of it as a little square of

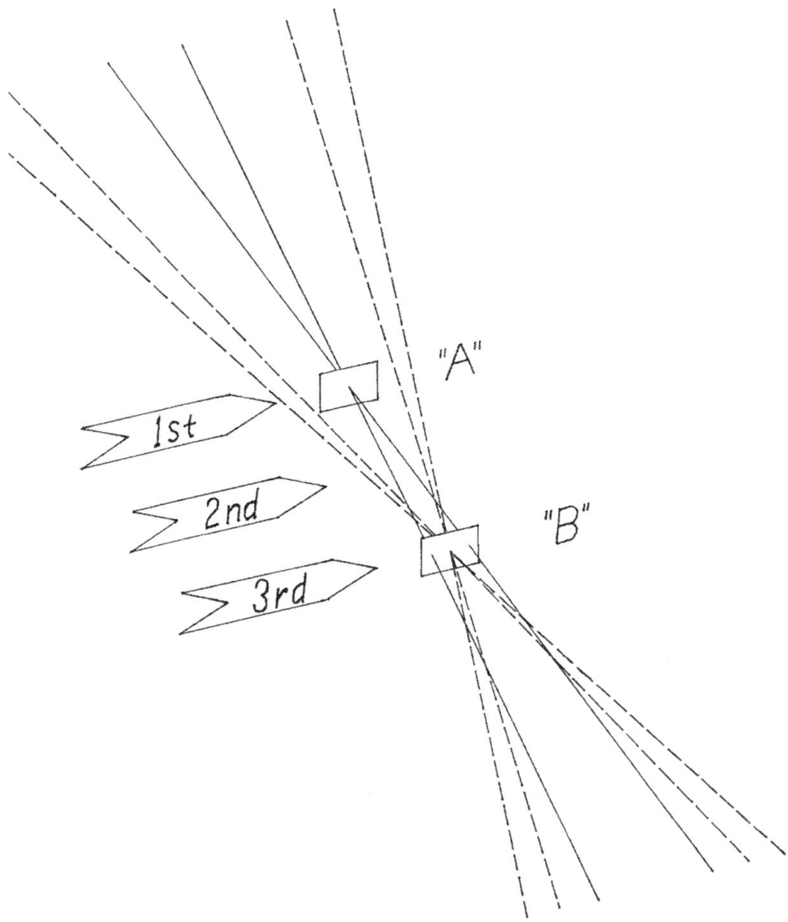

Figure 5a.

film) to represent each component's center of curvature. We have located three "arrows", labeled 1st, 2nd, and 3rd for three positions along the optical axis. We're going to explore the optical axis in this region with the KE, and inspect the mirror with it positioned, in turn, from each of these locations represented by the little arrows.

We show these locations identified by arrows to remind us that in each location we will bring the KE in from the left, starting with it well clear of any returning light rays from the mirror, so that we may observe the order of progression of the unfolding appearances as the KE is moved inwards. We will omit any depiction of our "imaginary" light source at each center of curvature to keep the drawing uncluttered. And this time we will not show the knife edge, either, leaving you to imagine it and its action as it moves inwards from each of these locations in turn. And lastly, we will remove the little squares of ground glass out of the way, now that we know where both centers of curvature actually are.

Let's start with the KE in the position marked 1st, the position closest to the mirror, and work successively outwards to the other locations, in turn.

As we begin moving the KE in from the left, it first encounters light rays returning from the left half of the outside annulus of the mirror, obstructing its left-most rays first. As the KE continues slowly advancing inwards, more and more light from the annulus is obstructed, progressively from left to right, and we see a very dark shadow proceeding inwards across the annulus from left to right. The shadow, so far, is moving in the same direction as the KE. By the time the KE is nearly just adjacent the optical axis, the left half of the mirror's outside annulus is almost completely filled in with dark shadow. Finally, as the KE moves the very last, tiny increment of distance to bring its edge just to the OA, it begins to partially and *simultaneously* obstruct light from every part of the central component of the mirror here where it crosses the optical axis.

As the KE obstructs this light, the central component grays out, or "nulls". We have detected the central component's center of curvature with the KE. Halting the KE's advance now, we note the mirror's overall appearance, and it appears as in Figure 5b, with the left half of the outside annulus completely dark, its returning light completely obstructed by the KE. The central region is nulled, with its light only about fifty percent obstructed. The right half of the outer annulus is still completely bright, as all of its rays pass well clear of the KE's edge, not at all obstructed in the least.

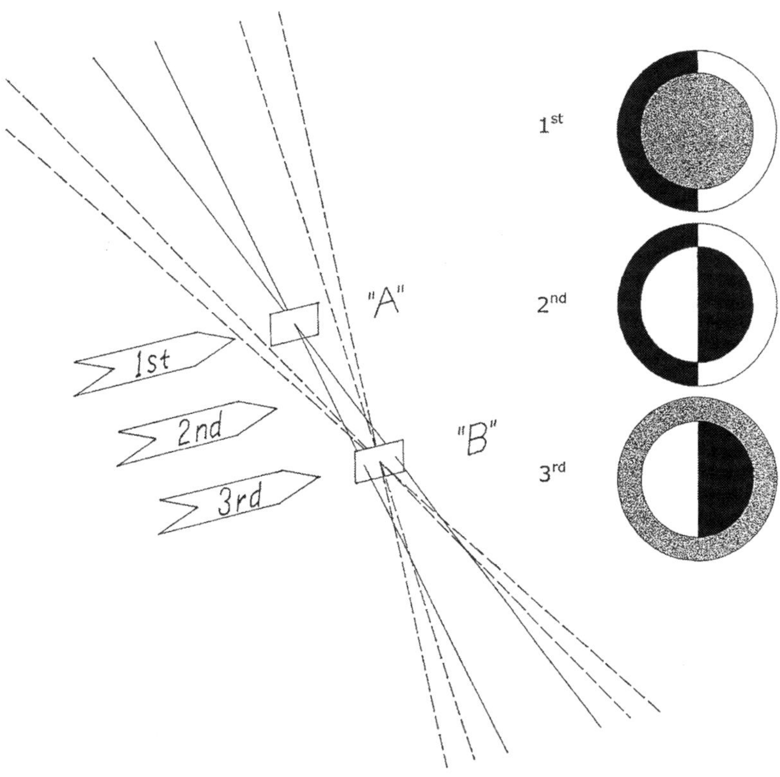

Figure 5b.

After taking careful note of the mirror's appearance with the KE in its last position at the arrow marked 1st, we next withdraw it laterally, away from the OA until no light from the mirror is obstructed, and then back it away from the mirror until it is at the arrow marked 2nd, midway between the centers of curvature of both the mirror's components. With one's eye again in place on the optical axis looking at the mirror, the KE is again brought very slowly in from the left until it begins to obstruct the leftmost rays of light returning from the mirror. As in the 1st position, the leftmost regions of the outside annulus begin to darken first as we see the KE's shadow again move in from the left, in the same direction as the motion of the advancing KE.

Continuing the slow, rightward motion of the KE, we begin to get close to the OA and the left-hand side of the outside annulus becomes nearly filled with dark shadow as before when we were working at the 1st position closer to the mirror. But then, a curious thing begins to happen: as the KE continues its slow advance up to the OA *a dark shadow begins to move in from the right, traveling leftwards across the right half of the central component of the mirror.* The rightmost regions of this component begin to darken as the KE obstructs their reflected light on the left side of the OA where they've crossed over a little ways after having passed through C of C for this central component of the mirror.

Continuing the KE's advance right up to the edge of the OA we observe this shadow continue and complete its advance inwards from the right hand side of the central component until the entire right half of this component is in dark shadow. After stopping the KE's advance right at the OA, we note the overall appearance of the mirror. The light from the left half of the outside annulus is completely blocked, and the light from the right half of the inside component is completely blocked, and both of these areas appear very dark. However, the returning light from the right half of the outside annulus passes well clear of the KE and it remains brightly illuminated. In addition, the light from the left half of the inside component passes well clear of the KE and this area also remains fully illuminated. The overall appearance of the mirror is as in Figure 5b, "2nd".

5th Essential Foundation Lesson:

We will finally relocate the KE in the 3rd position, starting again with it well clear of all returning light rays and then advancing it in towards the OA from the left. As it advances in from the left, the first light it encounters and obstructs is that returning from the right half of the inside component of the mirror. This light has already crossed over the OA and is well clear of it to the left in this 3rd position. As the KE continues its inward advance towards the OA, the edge of the dark shadow advancing in across the right half of the central component moves leftwards, in the opposite direction of the KE's motion until the right half of the inside component is nearly all dark. Then, as the KE advances the last small increment up to the OA, it begins to obstruct the returning light rays from every portion of the surface of the outside annulus *equally and simultaneously,* causing this outer annulus to gray out, or *null all over simultaneously.* The KE has detected the C of C for this outside annulus of the mirror. We halt the KE's advance here, right at the optical axis, and survey the mirror, noting its appearance, as in Figure 5b, "3rd". The right half of the inside component is completely dark, all its returning light occulted, and the left half is completely bright, none of its returning light occulted. The outside annulus is evenly nulled, as our KE is precisely at its center of curvature.

Exploring the optical axis of our "Lord Rosse Special" with the knife edge has been our first experience in surveying the surface of a concave mirror whose figure is something other than a simple sphere. We've garnered many different lessons with this exercise that will now be well understood intuitively, without further necessity to elaborate verbally. One, however, needs to be gotten in mind in a very unequivocal fashion, and so we now state it here as:

> **Essential Foundation Lesson #5:** ***The appearance of any concave mirror when surveyed with the knife edge will reveal the deviations of the surface from a true sphere as the knife edge is moved along the optical axis.***

Correcting Spherical Aberration

Before leaving our "Lord Rosse Special" it will be instructive to understand why it didn't work so well as a remedy for spherical aberration.

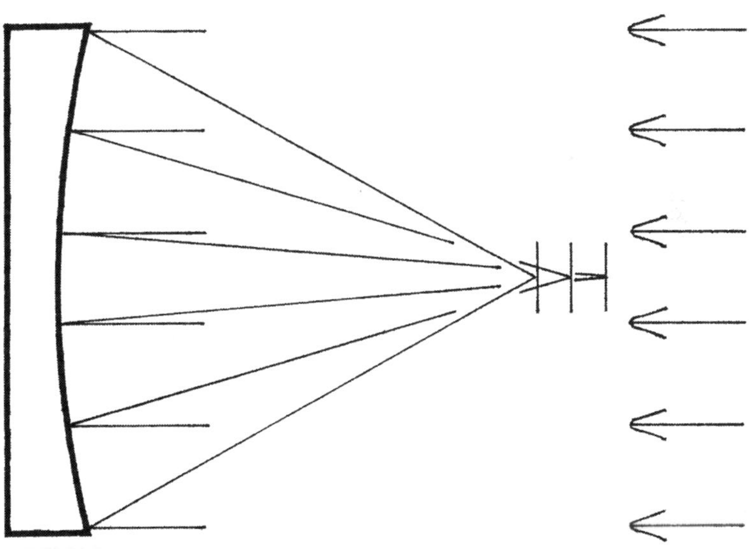

Figure 6a.

In Figure 6a we show a very short focus mirror in cross section receiving light rays from infinity and focusing them along a short region of its OA, as in our earlier diagram, Figure 2a. In that previous diagram we showed light rays from infinity striking the mirror near its edge and near its center and being reflected onto a short region of the OA. For Figure 6a, however, we show some rays in addition to those for the central and near edge regions. These rays are striking the mirror's face in a zone intermediate between its edge and center regions. The illustration makes it plain that the spherical mirror cannot focus rays of light reflected from any of these zones into the same focal plane. Three short lines lying across the OA represent the three different focal planes for light reflected from these three zones or areas of the mirror.

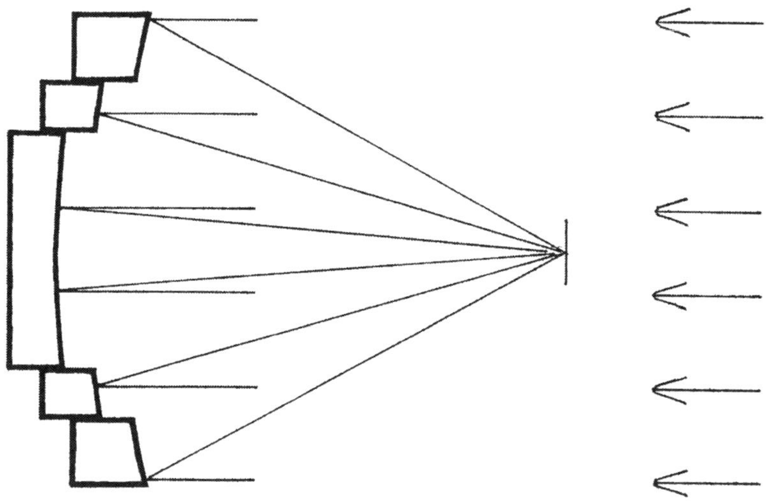

Figure 6b.

Contemplating this diagram should lead one to an intuitive insight: for as many zones as we care to demarcate the mirror's concave surface into there will be as many disparate focal planes for. It should now be clear why a two component "Lord Rosse Special"

will not work well; it is optimized for only two zones for the entire mirror: a narrow zone near the edge, and a small region very near the center. All other zones for either component of the two component special will still have widely disparate focal planes.
Figure 6b shows a hypothetical three component mirror in cross section. We might reasonably expect this three component mirror to work better than a two component one, focusing reflected light from the zone intermediate between its edge and center into a focal plane more congruent with those for focused light from the other zones. But a little thought here will reveal the impossibility of such an approach: each of the three new components is still a section of a sphere, and therefore will each have disparate focal planes for its different zones. In order to optimize the correction to bring the many disparate focal planes congruent, we would have to go on "surgically" subdividing the mirror further and further into more and more nested components, annuli, offsetting each one by the required amount from its neighbor along the OA.

6th Essential Foundation Lesson:

As a conceptual exercise, however, Lord Rosse's approach points the way to a more practical solution: we can offset the successively more central regions of the mirror, successively, from the successively nearer edge regions (relative to a sphere) simply by excavating these successively nearer central regions more deeply. We will then no longer have a concave surface representing a revolution of a circle (a sphere) but some other species of figure of revolution. And this is in essence the time-tested method for altering a spherical mirror's figure into one for which all zones of its concave surface have congruent focal planes, i.e., one that will focus light from its entire surface into the same focal plane. The particular species of surface of revolution that can do this is a curve that geometers long ago designated as one of a family of specifically defined sections of a cone: the paraboloid. And now we are well enough prepared with previously learned founding concepts to introduce the very important:

> **Essential Foundation Lesson #6:** ***The properties of the sphere and the paraboloid are exactly reversed with respect to each other with regards to each's conjugate focal planes. The sphere, while unable to focus light from infinity into one focal plane, will exactly focus all light received from its center of radius of curvature into one focal plane. Alternatively, the paraboloid, while unable to focus all light from its center of curvature into one focal plane, will exactly focus all light received from infinity into one focal plane.***

Strictly speaking, a paraboloid does not have a center of radius of curvature, since it is not a sphere; but we will take liberty and use language loosely here in order to help illustrate a concept.

This curious order of features of the curves of conic sections seems almost to have been by design for those of us who want a convenient and quantitatively precise test procedure for monitoring the development of the paraboloid during figuring. In particular, we are happy that the paraboloid cannot return light from a source located at its approximate center of curvature into one focal plane, but instead returns it into different focal planes strung out along its optical axis.

Summary, in Graphic

We're ready to learn the actual test procedure now. Since the essence of the test is embodied in our last essential foundation lesson, we will briefly recap this lesson in a compact, graphical form, so that you will get it firmly in mind. Let's look at Figure 7.

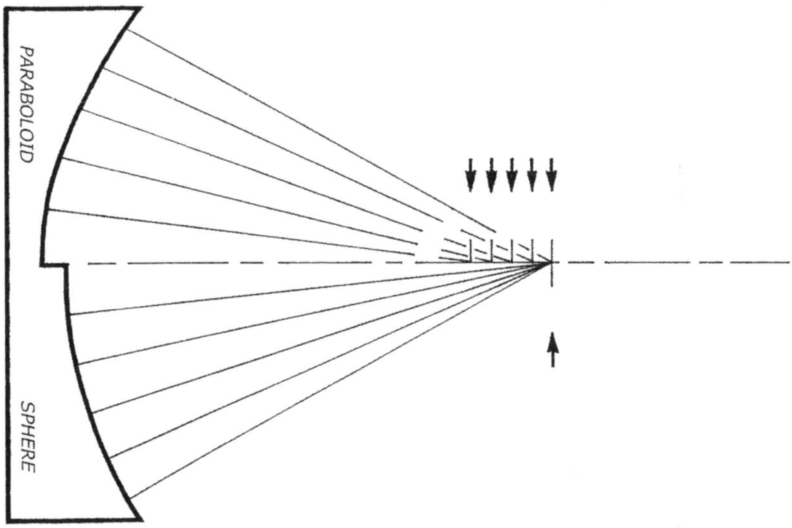

Figure 7.

We show yet another, curious, hypothetical mirror in cross section here. This imaginary mirror is divided into two different regions: the bottom half has been left spherical, but the top half has been figured paraboloidal. On the optical axis (out to the right) we've located the center of curvature of the lower, spherical half of the mirror. Light rays are fanning out from this C of C, striking the mirror, and returning back to its C of C. The little arrow below the optical axis marks the C of C.

Now, the upper half of this mirror is paraboloidal. We may reference it with the lower, spherical half of the mirror by specifying that it's very edge zone or region is *congruent with the sphere represented by its lower half.* This illustration makes clear, at a glance, what we used words to describe as our last essential foundation lesson: the sphere can return all light originating at its C of C precisely back to that C of C; the paraboloid cannot. For this paraboloidal part of the mirror, we show light rays emanating from the C of C of the lower, spherical part its edge is congruent with. Note that it can return only light from this edge zone back to the sphere's C of C. All regions, zone wise successively closer to the center of this paraboloid, will return light to focal planes that are successively closer to the mirror. The five different arrows pointing downwards indicate the five different focal planes for light reflected from the five different zones on the paraboloid (the same zones, radius wise, as for the sphere). All that remains is to show you that the disparity between these various focal planes for the paraboloid *can be specified* for the figure we desire, and that *they can be commanded into their desired locations along the OA by figuring the mirror.*

Now, although not critically necessary, it will be useful to understand an interesting lesson here before we begin testing.

In the course of testing their mirrors, experienced users of Foucault more or less pretend, when observing the knife-edge null a narrow zone on the mirror, that they have detected this narrow zone's radius of curvature. It should now be clear to all, however, that this notion can only be a useful fiction for the mirror maker. *No zone located on a paraboloid, however narrow, is spherical, and therefore cannot have a center of curvature.* But since a very narrow zone can return most of its light to a very precisely detectable center on the OA, we tend to think of it as in some way approximately spherical.

Foucault Test Apparatus

The purpose here is to learn the test *procedure*, so we will limit our discussion of **tester** features to essentials. Our illustrations so far contain a mysterious "perfect" light source. We only show for example in Figures 3a-c and 5b the knife-edge. The light source and slit are never shown. In the real world our test apparatus is comprised of three basic functional components: (1) A mounting platform stage providing linear, translational motion in X and Y axes; (2) the knife-edge (KE); and (3) a very minute light source behind a slit.

The knife edge, light source and slit must be carried on this translating stage in a plane perpendicular to our mirror's optical axis. The mounting stage Y axis is set to move the KE-slit combination, via a micrometer or dial screw along an axis ***parallel*** to the OA of the mirror. It is critical the Y axis (OA) movements be resolvable (measurable) to at least 0.001 inches. Resolving to 0.0005" is even better. The X axis is set to move the KE-slit combination laterally, at a right angle to the mirror's OA. The X axis can be adjusted with a screw, to move the KE-slit combination into and out of the C of C, but does not need to be measurable.

We will design a translation stage which will hold both the knife edge, and the slit together. The arrangement can be envisioned as "one long slit with half the slit missing on top." See Figure 8a. To form the KE and slit, a very fine, sharp edge is desired so as not to project diffraction artifacts or shadows from rough edges onto the mirror surface as viewed from the eye.

A ready-made fine, sharp edge is available with any number of common, commercially available razor blades – the type used for shaving. The best choice among these, however is the "injector" type *single-edge* blades. A single edge blade is not only safer to handle, but a one-sided bevel on the sharp edge is preferred. (This is not the general case as many single-edge blades are beveled to a

"V" shaped edge. If possible, obtain the type of blade which is beveled on only one side.)

The knife-edge and the light source are both mounted congruently in a plane (mounted in the same plane) through which the mirror's OA passes perpendicularly. Figure 8a illustrates only the simplest construct, simply to convey the idea of the "knife edge" and the "slit" in the same assembly. The positions of the razor blades in Figure 8a are shown as they would look, not mounted, and from the "eye side". The slit should be adjustable in width if your skills permit.

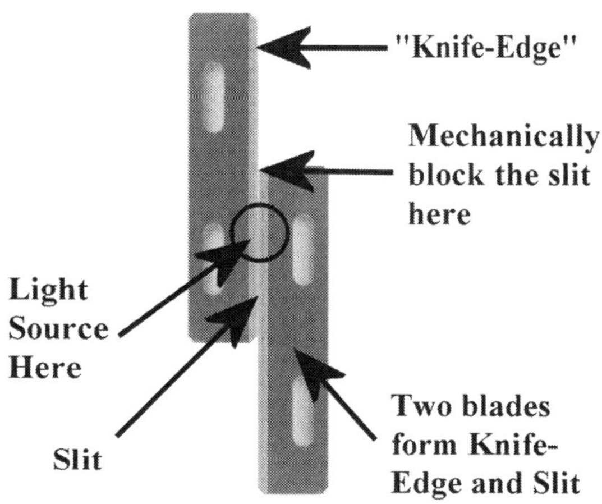

Figure 8a.

The most important part is to fashion one edge of one razor blade to be the knife-edge while at the same time be one jaw of the slit formed by the other blade. Thus, if the blades are staggered, as shown in Figure 8a, by one half their length on the mechanism that holds them, you will never see any "false astigmatism" as you would when doing a fast mirror up close with a side by side arrangement (i.e. the KE and slit side-by-side).

Also, if the light is fixed, and only the knife is permitted to move along the OA, you are never really at the center of curvature for any zone but only approximately; the differences of course are small, but however small the error, if you are not testing at center of curvature, always relative to the same point on the mirror, you are not presented with a true representation of the situation.

But when the knife edge and light move together, they always end up exactly at the C of C of the zone under test. There is a tremendous additional benefit to be gained by this arrangement. More on this later.

Let us cement in our minds what the KE-slit assembly should look like. In Figure 8b, we are now looking at the razors from the obverse of Figure 8a. The razors are shown mounted on a heavy cardstock or a thin aluminum plate which will be mounted to the X-Y translation stage described above. The view now is from the mirror side.

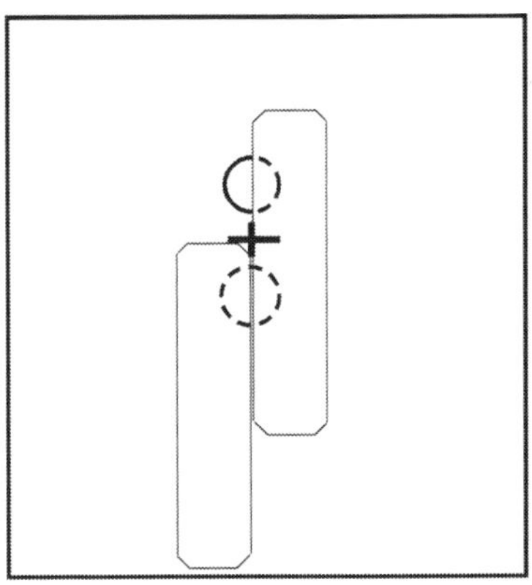

Figure 8b.

Two holes are drilled in the cardstock or plate, as close together as practical, yet which still allows for the placement of the razor blades, and the lamp assembly. In the illustration, we are viewing the holes *through* the razor blades, hence the holes are shown as dashed as they are covered by the blades. The lamp would be installed to illuminate the slit through the lower hole (from the other side), while the KE would be viewed through the upper hole (also from the other side.) The "+" in the figure, shown between the two holes indicates where the OA will fall when the KE-slit assembly is in position for a test. The holes, though shown as relatively small in the figure, could each be 5/16" or even as large as 3/8".

As mentioned above this knife-edge/light assembly is in turn mounted on a <u>moveable</u> platform stage so it can be moved at <u>right angle</u> to and also <u>along</u> (parallel to) the mirror's OA. A dial or screw micrometer is provided for reading the amount of travel in the Y direction (motion along or parallel to the OA). *Inasmuch as our light source and knife-edge are both mounted on the same plate carried on the platform stage*, they <u>move together</u> as a unit in both X and Y axes.

You will also find it valuable to design the mechanism holding the razor blades such that the slit formed, can be adjusted. *Slit widths* from 0.01" to as small as 0.0005" are desirable for the most contrast for each zone under test, and for different diameter mirrors. A very narrow slit, of course demands a variable lamp brightness to compensate!

Now to discuss the benefit of this KE-slit arrangement.

We have described a combined KE-slit arrangement such that, as you turn the screw for the X axis, the KE-slit combination can be moved into or out of the OA at the C of C. The location of the KE-slit assembly along the Y axis is known by the reading on the micrometer or dial screw providing 0.0005" accuracy.

Assuming a left-to-right motion in the X axis, the mirror forms an image of the illuminated slit which, at first, is further to the right, away from the KE. As the screw is turned, slowly advancing the KE-slit toward the OA, this slit image slowly converges toward the KE. Refer to Figure 8c.

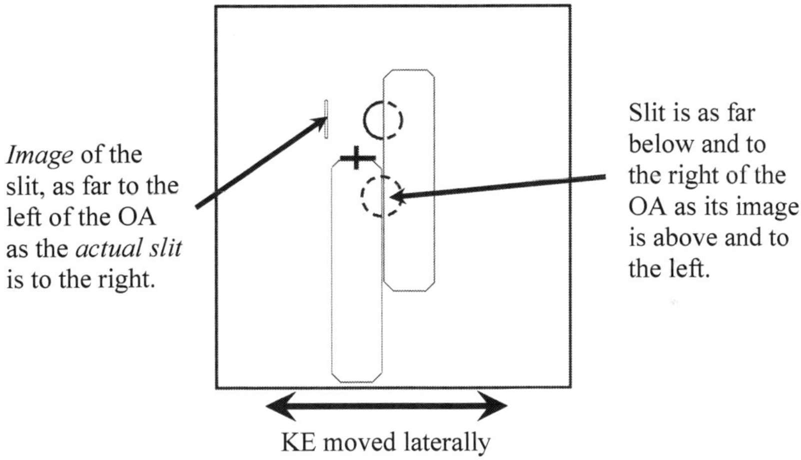

Figure 8c.

Now imagine you are turning the screw to advance the KE (*to the left* in Figure 8c). The inverted *image* of the slit will move *in the opposite direction.* The edge of slit image, due to its width will reach the plane of the OA *slightly before* the KE.

Ultimately, when the KE is right on the plane passing through the OA, as in Figure 8b, the other edge of slit image will **meet** the KE, and at this moment the edge of the slit and the KE are both at the C of C for that particular zone on the mirror. This configuration insures that not only the **slit** is at the C of C, and the **KE** as well, but also means that the edge of ***image*** of the slit falls on the KE at the same moment both the KE and slit reach the plane of the OA. In this manner, the tester *behaves* as though it has no slit as the other side of the slit image, the side which reached the KE first, has no effect on the KE.

The great benefit of this "slitless" behavior is that the slit can be comparatively *wide* without impacting the contrast of shadow at the KE. The design, as described above, of the KE-slit combination, accommodating an adjustable slit-width will permit the slit to be set as great as 0.01" wide providing a bright illumination yet simultaneously a high contrast shadow.

Surveying and Measuring the Paraboloid

Figure 9a through 9f shows the appearance of a fully parabolized short focal length mirror for six different positions along its OA as viewed with the KE. By convention we will always begin by pre-setting the micrometer for our tester's Y-axis movement at zero, and locating the tester to null the central region of the mirror. From there we will work the KE backwards along the OA away from the mirror, to find the null point, successively, for several different designated zones on the mirror. In Figure 9, under each depiction of the mirror's appearance ("apparition") for each setting of the KE, we show a drawing depicting the mirror's apparent cross-section. Remember, we said we would always think of a concave spherical mirror as *flat* when viewed as nulled from its center of curvature. Similarly, we will think of the shape of the paraboloid when viewed with the KE as a variation from the *flatness* of our "flat" *reference sphere.*

After nulling the very central region of the mirror, we advance the KE away from the mirror and stop at the location shown at Figure 9b. Note that the mirror appears to have an annular, circular "crest" surmounting an apparent, gentle bulge all around its center just a little ways out. Our KE is exactly at the C of C of a very narrow zone surmounting this crest. More accurately (as no zone on a paraboloid

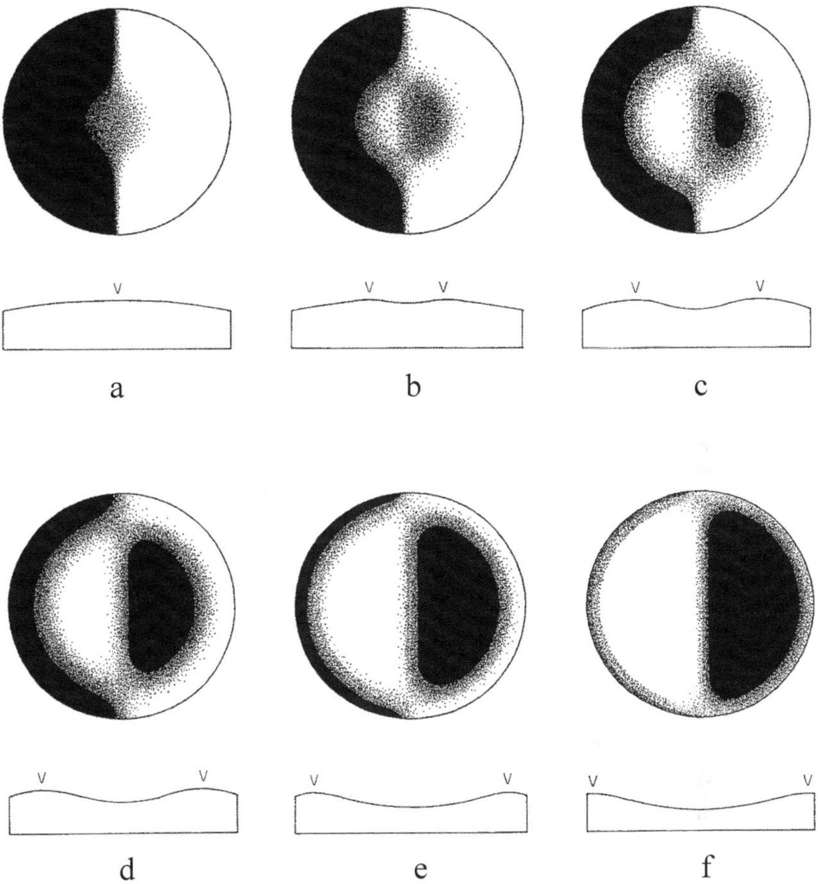

FOLLOWING THE CREST OF THE DOUGHNUT
If we start with the KE at the C of C for the small central area of the mirror and then advance it backwards along the OA smoothly, we will see the zone being nulled at any given location for the KE appear as a smoothly outwards rolling "crest" of a doughnut-like shape.

Figure 9 a through f.

can truly have a center of curvature) we are at that point on the OA where that zone's rays are exactly crossing it. Our micrometer will show us, when we inspect it, how far the KE moved backwards to provide this particular apparition of the mirror. The micrometer indicator will show us the KE's location along the OA where this zone's light rays cross over it, relative to its previous location.

We may continue backing the KE away from the mirror, noting, in succession, the other apparitions at c, d, e, and f. The micrometer will always show us the relative location along the OA for the C of C of the narrow zone represented by the crest of the bulge. In addition to being able to locate the C of C for any zone being nulled by our tester fairly precisely along the OA, we can also measure the location of the zone itself on the mirror, its radius from the center of the mirror. *And these are the only two quantities we need to determine accurately during the figuring of our mirror in order to shape it into a section of the true paraboloid.*

We will pre-determine which zones' centers of curvature we want to monitor before we begin figuring. Conventions or rules about the number and locations of zones for testing vary. The popular convention of dividing the mirror into zones of equal area probably is most advantageous. Zones of equal area will provide for increasingly narrower and more closely bunched zones, successively, outwards towards the mirror's edge. This seems reasonable in that we must figure the outer zones to tighter tolerances than the inner zones. It is very much true, as an older master once told me, that: "The edge zone sets the mirror's performance."

We will compute 5 zones of equal area. The algebra to determine the radius of the center of 5 zones of equal area is not difficult, but tedious, so only the results are presented here:

> The radius, r_1 of Zone 1 will always be 0.316 times the mirror radius, r_m.
>
> The radius, r_2 of Zone 2 will always be 0.548 times the mirror radius, r_m.

The radius, r_3 of Zone 3 will always be 0.707 times the mirror radius, r_m.

The radius, r_4 of Zone 4will always be 0.837 times the mirror radius, r_m.

The radius, r_5 of Zone 5 will always be 0.945 times the mirror radius, r_m.

Let use as an example a project to figure a ten-inch mirror of sixty inches focal length. To find the location of the middle of each zone (as a radius from the center of the mirror) for any diameter mirror, multiply the mirror's radius (r_m, in this case, 5 inches) successively by: 0.316; 0.548; 0.707; 0.837; and 0.945. For our **ten inch** mirror therefore, the middle of each zone computed in this way will, be, successively: 1.58"; 2.74"; 3.53"; 4.185"; and 4.725" as measured from the mirror's center (i.e., as radii).

For the mirror's curve to be the correct section of a true paraboloid for its given diameter and focal length, the *difference* of the C of C of each zone is fixed by formula, Figure 10. Stated another way, the location of each zone's C of C is farther away from the C of C of the very central region of the mirror by the value "d" of Figure 10. (Thus "d" for the exact center of the mirror is zero, and represents the reference point.) Therefore, for our 10", 60" focus mirror, "d" for each zone is found to be:

Zone 1 (0.316 r_m or r_1 = 1.58"): d = 0.010"
Zone 2 (0.548 r_m or r_2 = 2.74"): d = 0.031"
Zone 3 (0.707 r_m or r_3 = 3.53"): d = 0.052"
Zone 4 (0.837 r_m or r_4 = 4.185"): d = 0.073"
Zone 5 (0.945 r_m or r_5 = 4.725"): d = 0.093"

These values are determined by formula (Figure 10) where "r" represents the radius of a zone from the center of the mirror (r_1, r_2, etc.) and "R" represents the radius of curvature of the mirror (as imagined, of course, as spherical, before figuring). Note the radius of curvature, R, is twice the focal length, thus for our 60" focal length, 10" diameter mirror, R = 120". This is not quite the formula most experienced workers are familiar with, as more commonly their testers have their light source fixed and only move the knife edge along the mirror's OA. As explained previously, we will carry both the KE and the light source on a small plate together so that we may move them simultaneously along the OA. Thus the formula depicted in Figure 10 is exact for our method.

$$d = r^2/2R$$

Figure 10.

It is important to note that values calculated in the example above is *only good* for our example of a 10" diameter mirror of radius of curvature 120". The value "d" computed above will be close to, but nonetheless vastly different for another mirror even of only a slightly different radius of curvature. One must take care to compute the Zones, and the "d" values for each zone for the particular mirror in each case. Remember the case of the misfigured Hubble Space Telescope mirror!

Locating the Mirror's Zones

We need a practical method for accurately locating any given zone on the mirror for nulling with the tester's knife-edge. Let's look at Figure 9d again. This illustration depicts the number three zone (3.53"r) being nulled by our tester's knife edge. This zone divides the mirror into two equal areas, and is by convention referred to as the ".707 zone". How can we be sure that the area being nulled (equally gray all the way around- the gray "crest" of the bulge) is actually centered on the 3.53"r zone? *We will put a specially prepared marker in front of the mirror for locating its zones.*

Testing Masks

Zone locating masks for Foucault testing are of two basic types. The traditional type has two equal sized apertures cut into the mask for the left and right side of each zone. Over the years I evolved some major improvements in their design and application that improved their accuracy and convenience in use. Finally, though, I discovered the advantages of the "Everest" style zone locating mask, and began to make and use this type exclusively, rapidly incorporating improvements in Everest's basic concept just as I had with the traditional zone locating masks.

Everest's basic approach was to hang a section of yardstick in front of the mirror being tested with pairs of straight pins protruding from one edge to mark the radii of zones for testing. The straight pins would be seen in sharp silhouette against the zone being nulled- one could see their outlines, each on either side of the mirror, against the crest of the "doughnut" behind them. As embodied by Everest, the test is somewhat hampered by a perceptual defect. I have noted this defect and have improved the design by making each pair of straight, vertically standing markers (Everest's "pins") into *markers curved to the same radii as the zones they represent.* The improvement in certainty when locating a zone with this kind of mask is dramatic.

An example of this kind of mask is shown in Figure 11. In this particular example (an early form of my improved design) the little marker "horns" protrude up from the crosspiece that supports them. This earlier example has the pointed tips of the indicator horns lying along a meridian across the mirror's horizontal diameter. Horns about twice as long as these, extending equally above and below the mirror's meridian of horizontal diameter, are even better. These

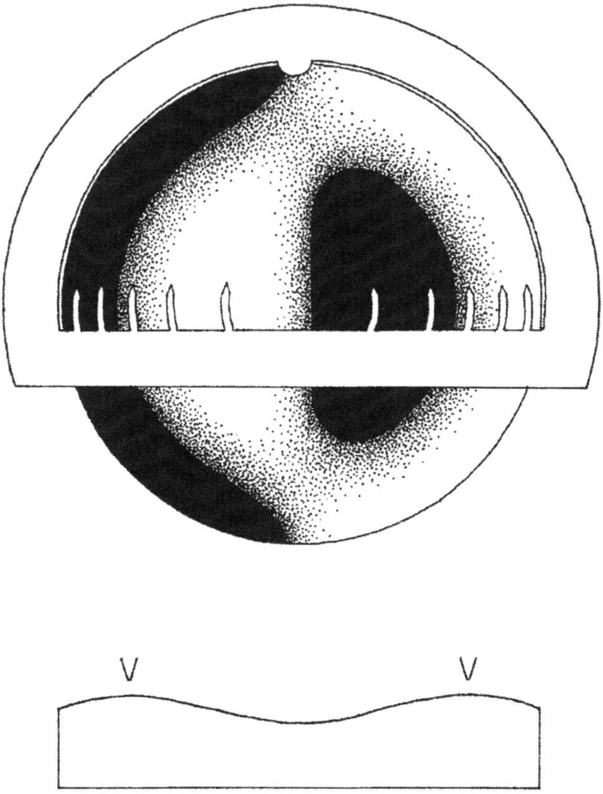

LOCATING ZONES

An "Everest" style (modified) zonal testing mask just in front of the mirror will allow us to position the KE to null a zone marked by any pair of the mask's curved indicator points (after Harbour). Above, we see the third pair of points in silhouette against the crest of the "doughnut", indicating that the KE is at that zone's approximate C of C. In this case, the KE is nulling the .7071 "r" zone of the mirror under test.

Figure 11.

curved indicator horns can be made quite long, since they accurately locate a zone *lying everywhere underneath each horn's entire length.* A curious perceptual effect is at work here. The longer the horns, the more certain the impression of the crest's location underneath them. When you make your first zone locating masks, make these horns as long as you please, *but each of them must be curved along its entire length to the radius of the zone it is intended to mark.*

The mask is easily prepared with a beam compass on poster or illustration board, and then cut out with a sharp hobby knife. The configuration shown in Figure 11 is just about perfect- but extend the narrow, curved horns upwards through the mirror's middle, horizontal diameter farther than shown. I have gotten best results with masks that have the horns extending an equal distance above and below the mirror's horizontal diameter. They should be kept quite narrow, especially for smaller mirrors.

Figure 11 shows the .707r zone (Zone 3) being nulled. The middle pair of indicator horns (third pair, outwards from mirror's center) appears to be lying directly atop the crest of the torus-like or doughnut-like bulge. We can have confidence with this indication that the KE is very close to the C of C of this zone. The appearance will be the same for the other zones represented by the other indicator horns when the KE is at their respective centers of curvature. In each case, that zone's particular indicator horns will appear to be lying directly atop the crest of the bulge.

Allowable Errors

As it turns out, figuring the mirror so accurately that the readings for the KE's positions along the optical axis fall precisely as predetermined is neither possible nor necessary. There are two reasons for this. First, there will always be at least a very small domain of ambiguity for the position of the KE when we try to null a zone with the KE on the optical axis. This is because the C of C of any zone being considered, no matter how narrow we define the zone as, does not truly lie on the OA. Second, the physical properties of light also decree a range of ambiguity in the location of the plane of focus for any given bundle of rays of light being focused into a point in the focal plane. In fact, no lens or mirror can actually focus light into an infinitesimally small point in its focal plane. Rather, when examined up close, we find the tip of the cone of a focused bundle of light not to be a tiny sharp point, but rather a very small *disk* with a measurable diameter. This little disk of light is the so-called Airy disk (after Sir George Airy, 7th Astronomer Royal, England). The Airy disk is sometimes also referred to as the "diffraction disk".

An image in the focal plane of any mirror or lens is an accumulation of tiny Airy disks all over its surface, representing the tips of many cones of focused light from many different points of origin in the object or field of view being imaged. Each of these myriad cones of focused light is a reflected bundle of parallel light from a single point source in the field of view of the telescope. Each entire bundle of parallel light represents each point source in the field of view and approaches the mirror or lens at a slightly different angle. Each of these bundles of light is then reflected (or transmitted through a lens) at an angle that corresponds to the angle it approached the lens or mirror. Consequently, each bundle of focused light places its Airy disk in a place in the focal plane that corresponds to its point of origin in the field of view. We may think of these little disks as image "pixels", somewhat analogous to the image sensing pixels in modern CCD camera arrays (though they do not sense anything but are actually picture elements themselves). Or an even better analogy might be to think of them as similar to the dots in a newspaper picture.

Now, the size of the Airy disks at the tips of each of these bundles of focused rays can be measured, and is different for different sized lenses or mirrors. The size of the Airy disk is a function of the focal ratio of the mirror or lens. If we move slightly inwards along the OA (towards the mirror) from one of these disks in the focal plane for a cone of focused light, we will finally come to a place along the cone where a cross section of it will be a disk having the same diameter as the Airy disk at its tip. Conversely, if we move outwards along the OA (farther away from the mirror or lens) from the Airy disk in the focal plane, we will again come to a point where the re-expanding cone of light has a circular cross section that again equals the diameter of the Airy disk. If we inserted a small square of finely ground glass in the focal plane and moved it back and forth through the focal plane between these two locations we would not see the little focused dot of light on the glass change diameter. In short, *it is quite impossible for us to find a precisely defined focal plane for any mirror or lens.* Rather, we will have this very short region in which the focus will be found to be acceptable. Thus, we require to figure our mirror only accurately enough that the cone of light focused by any given zone on the mirror will fall somewhere between these two locations along its optical axis. This range of locations for the C of C of any zone constitutes our allowed (tolerance) error for its location.

Tolerances

The amount of error that is allowed for the location of the focal plane to deviate from its ideal location for any given zone on a mirror has been worked out for us with the science of geometry. For our purposes it is not necessary to elucidate the entire method of determining the allowed error. Rather, we want to know how these allowed amounts of error translate into allowed ranges of location for centers of curvature of any given zone for our mirror. In other words, how large a range of position is allowed for the location of the KE for any given zone under test?

This range of allowed locations for the KE is determined by the simple formula in Figure 12. We will call this amount of allowed range of variation in the location the C of C for any zone "X". This

$$X = 2p\,R/r$$

Figure 12.

quantity, X, represents the amount of distance the KE may be closer to the mirror by, or farther away from the mirror by, than the computed ideal location of each zone's C of C. We show a summary of the meaning of X in illustration in Figure 13. In this diagram we see the cone of light returning from our tester's light source, focusing down to a near point in its focal plane located at its center of curvature. We have inserted a small square of ground glass in focal plane and note the tiny spot of light representing the Airy disk projected onto it. We may move the ground glass closer to the mirror by the amount "X", before the cross section of this cone of light represented by the spot projected onto its surface is larger than the Airy disk (position marked "1st"). Also we may move it farther away from the mirror, passing through the focal plane at C of C and advancing beyond it again by a distance equal to "X", (position marked "2nd") before the cross section of the re-expanding cone of light is again as large as the Airy disk.

For the other terms of the formula in Figure 12, "p" is the radius of the Airy disk at the mirror's focus for infinity, "R" is again the radius of curvature of the mirror and "r" is again the radius of the zone on the mirror under test.

To find "p", the radius of the Airy disk for any mirror at its focus, we will use the expression in Figure 14 (page 56), where "F" is the focal length, and "D" is the diameter of the mirror, and "w" is the wavelength of yellow-green light (.0000216") that has by convention been adopted as the standard for these purposes. After determining the radius of the Airy disk for our mirror, we can plug it into the formula as in Figure 12 and determine X, the allowed variation of location of C of C for any zone.

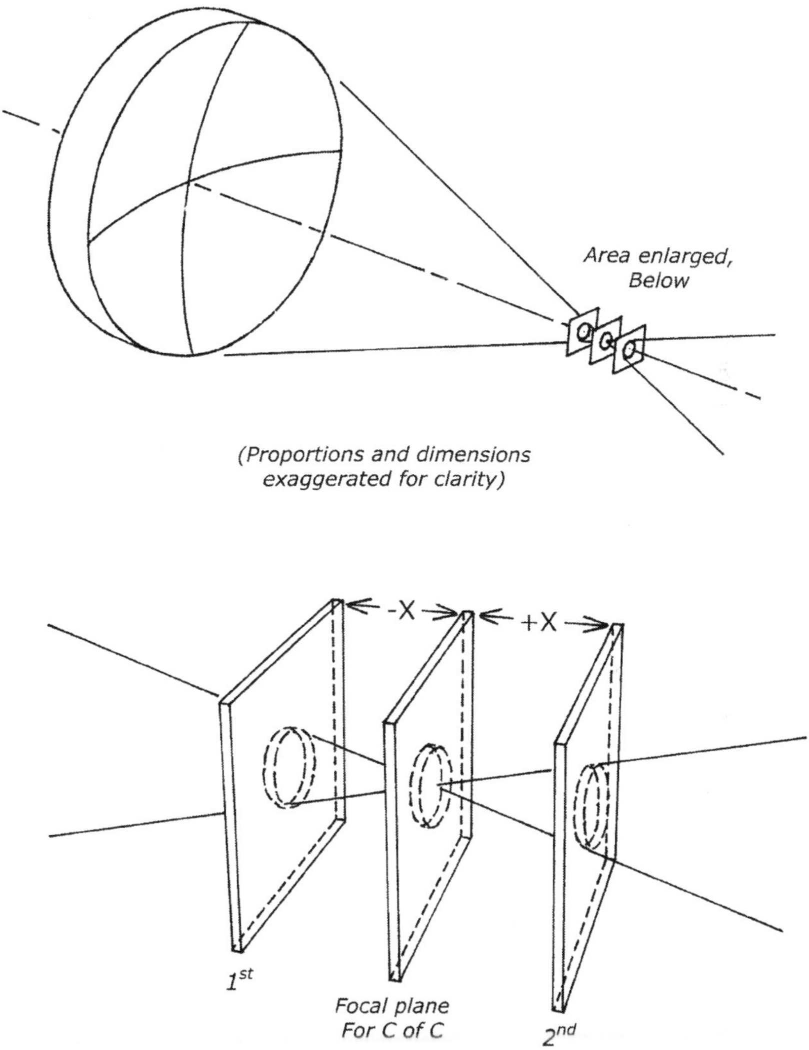
Area enlarged,
Below
(Proportions and dimensions
exaggerated for clarity)
-X
+X
1st
Focal plane
For C of C
2nd

Figure 13.

Now, we've already computed "d" for the five zones whose centers of curvature we wish to command into their predetermined locations on the OA through figuring. For each value of "d" for each zone, we add "X" to and subtract "X" from. Any reading for the location of the C of C for any zone that falls between these computed values is acceptable- with a certain caveat that we shall shortly stipulate.

$$p = 1.22 w F/D$$

Figure 14.

Interpreting Test Results

Figuring our mirror so that the centers of curvature of each zone as measured with the KE fall within tolerances will give us an acceptable mirror. However, using a graph to visualize the relationships of the plot of each KE setting with each other KE setting will help us to visualize and plan the best approach to refine and idealize the mirror's figure.

A graph of the values of "d", and "X", and the actual locations of C of C for each zone as measured with the KE is easy to construct. We show such a graph to help manage testing and figuring in Figure 15a. The vertical line on the left side of the graph has index marks in hundredths of an inch and represents the position of the KE settings along the OA. The horizontal line at the bottom of the graph represents the mirror from its center outwards, radius-wise; the vertical lines extending up from this line represent the locations of the five zones that we will test for, radius-wise from the center of the mirror.

The horn shaped figure sweeping upwards to the right and away from the center of the mirror represents the envelope or domain of allowed readings of the KE for the centers of curvature of any zone on the mirror under test. The middle curved line (inside the "horn") is for the pre-computed plots for "d" for any zone on the mirror (location of C of C relative to C of C for center of mirror). The upper curved line of the tolerance horn represents the allowed range of positions for "d" that are farther away from the mirror than the ideal positions. The bottom curved line of the tolerance horn represents the allowed range of positions for "d" that are closer to

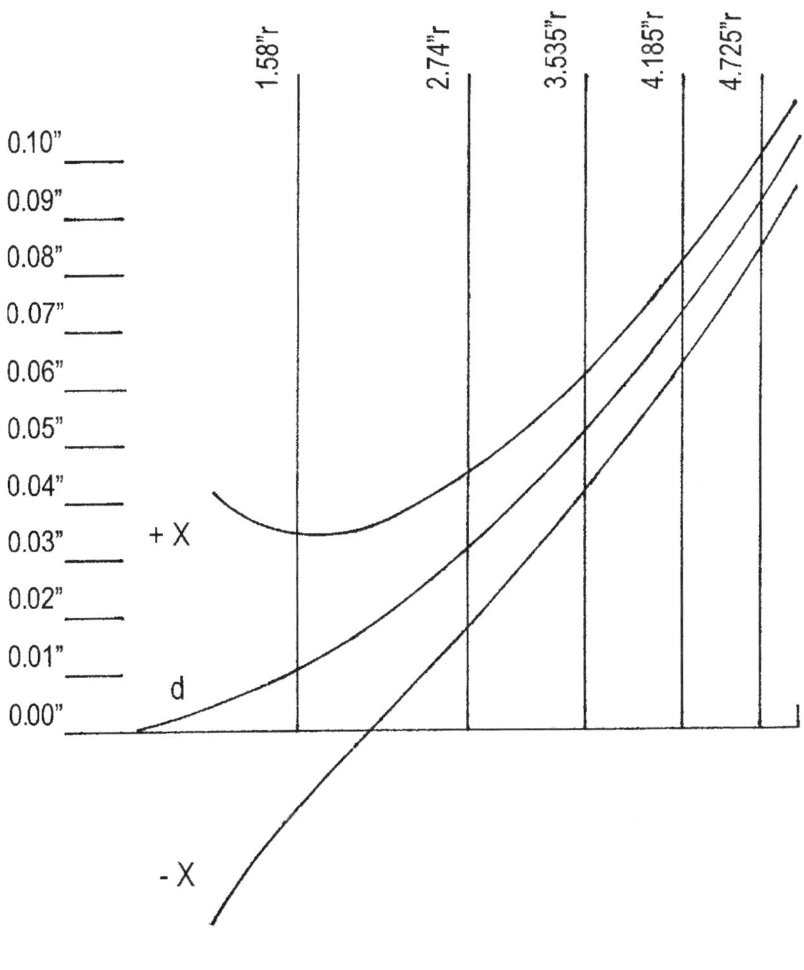

Figure 15a.

the mirror than the ideal positions. In order to plot relatively smooth and accurate lines for the values of "d" and "X", it is helpful to compute them for zones with radii in half inch increments for the mirror, even though we will be testing for only the five zones previously computed. Metric ruled graph paper is convenient for making these test result graphs, as the centimeter markings are a convenient size to represent hundredths of an inch, and they are subdivided into ten smaller units (millimeters) to help one represent thousandths of an inch. Use these to represent the vertical ordinate, for plotting the relative locations of the KE settings. For the horizontal, radius-wise ordinate extending to the right, use a ruler. An inexpensive machinist's ruler divided in tenths and hundredths of an inch is handy for this purpose.

Quick Summary: Procedure and Analysis

You now have everything you need to know to accurately test and plot your test results for the mirror you are figuring. To get everything concisely and compactly in mind, we will now summarize the test procedure and management of test date.

Set the "Y" axis stage of your tester to its zero setting, and carefully locate it along the mirror's optical axis to null its very central region. With your pre-cut zone testing mask in front of the mirror, back the Y axis stage carrying the KE away from the mirror to find the C of C of the first zone and note its location as indicated by the micrometer (write it down). Then, back the KE up again until the next zone as indicated by the horns on the mask is nulled and note, again, your micrometer's reading. Next, repeat the procedure for the third zone out from the center, the fourth, and finally the fifth, recording the location of each one's C of C as indicated by the micrometer. You will find during testing that unless the Y-axis stage runs truly along the mirror's OA you will have to manipulate the lateral, X-axis movement, to make a good null each time. This is okay.

Plot your recorded locations of each zone's C of C on your previously prepared graph for this purpose in their correct locations, and then connect these plots with lines as shown in our example in Figure 15b. Of course, at the beginning of figuring, the line of the KE settings will probably be "all over the place", not even

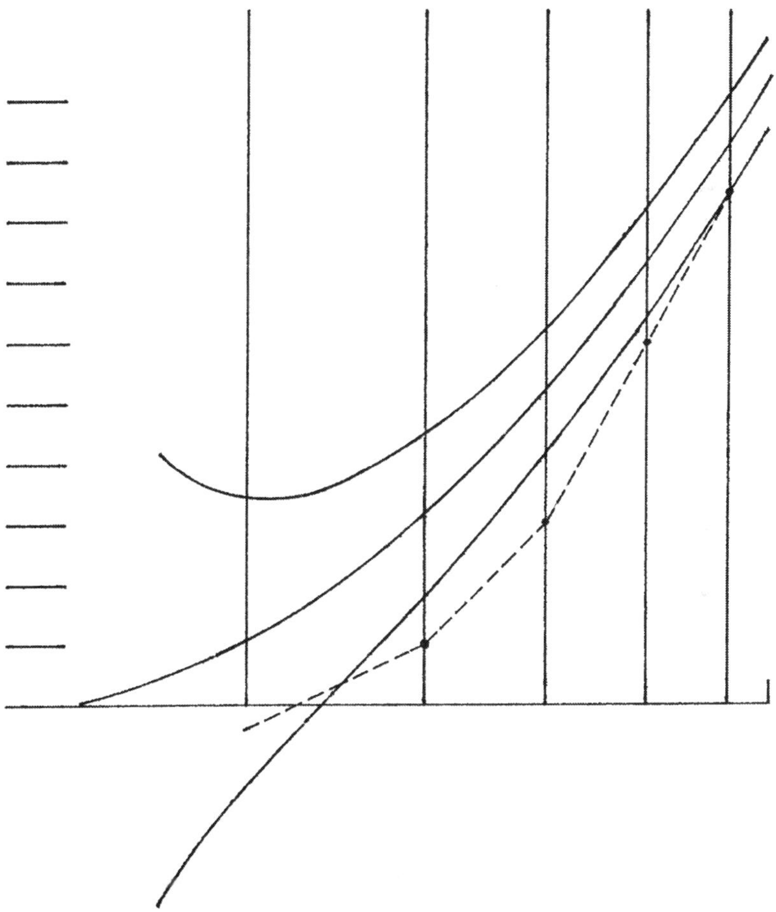

Figure 15b.

approximately fitting inside the tolerance horn of the graph. But, you might get a pleasant surprise: you might be "in the ballpark" from the start. I knew a gentleman who "accidentally" figured his mirror into a good paraboloid just by polishing it out! (Don't expect this.)

Let's consider Figure 15a as a representative example of typical KE settings for one test run somewhere near the end of the figuring process. Note that the first reading of the KE for the first zone is actually closer to the mirror than the C of C of its central region, i.e., we had to *advance the KE towards the mirror to find it,* rather than find it pleasantly located in its proper location a tiny ways away from the C of C of the central region. The plots for the second, third, and fourth zone fall outside the tolerance horn. But note that the overall shape of the connected plots approximates the shape of the tolerance horn, envelope.

In Figure 15c we have relocated all of the plots farther up on the graph by an equal amount, each of them, until they all fit inside the tolerance envelope. This is allowed: *it is merely the equivalent of starting with the tester located closer to the mirror by that amount of distance.*

Now the plots of all centers of curvature are lying everywhere inside of the tolerance envelope. Our ten-inch mirror of sixty inches' focal length (focal ratio of six to one, or "f/6") is now well enough figured that it will show no spherical aberration in use. Even if the tips of the cones of focused bundles of light from any zone on the mirror come to a focus into a plane farther away from their ideally computed ones, the blur circles representing the cross sections of these cones of light where they intersect their planes of ideal focus and pass through them will be no larger than their Airy disks at true focus would be in their ideal focal planes.

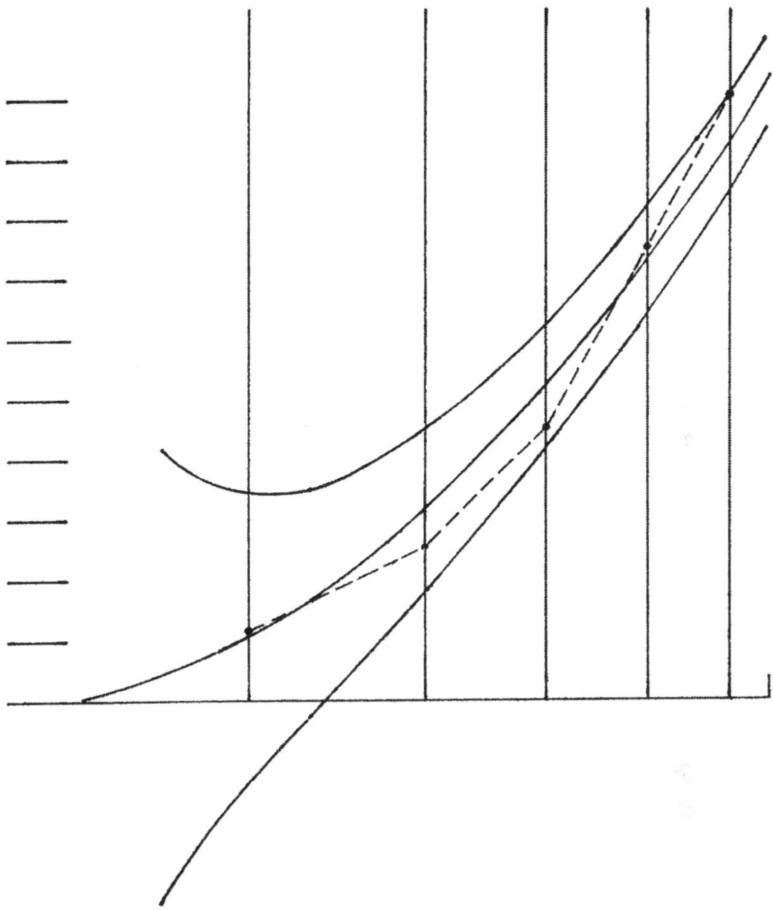

Figure 15c.

The tolerances as computed by the formulae given are considered "loose" by most authorities; that is to say, they are considered to be the least demanding for acceptable performance for a telescope's objective mirror, and many authorities recommend making a mirror's curve to at least twice as demanding tolerances. By all means, one may continue figuring until his or her mirror's plots of

KE settings fall very close to the middle curve of the graph (for values of "d").

I have tested many mirrors on the stars whose plots were spread out for the full tolerance envelope allowed by the graph. On nights of extremely steady air and at magnifications approaching 50X per inch (for larger mirrors) none of them *ever showed any detectable halo of spherical aberration.* However, the caveat we promised to

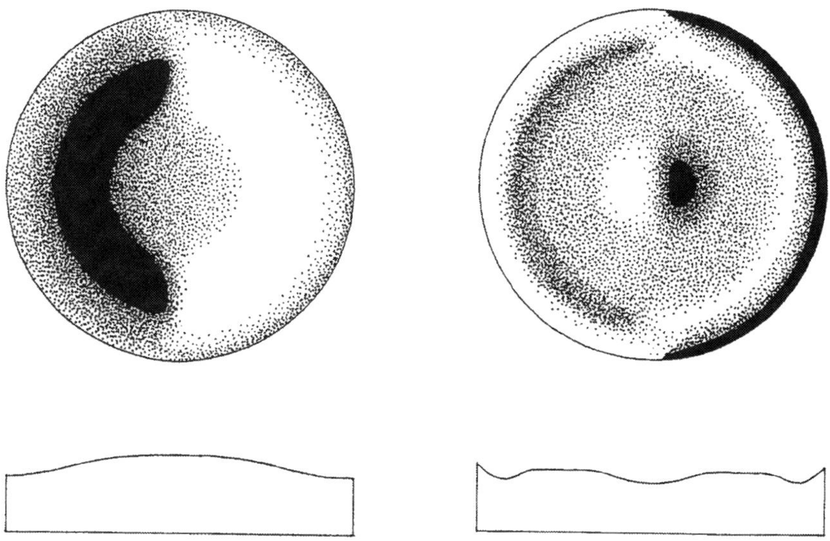

a. b.

ONE AND THE SAME

Note how radically different the figure of the mirror appears with the knife- edge set at two different locations along the optical axis. These two focographs represent the appearance of the very same mirror with an irregular, zoned figure from two different vantagepoints of the KE.

Figure 16 a and b.

convey to you in this regard (using up all the available space inside the tolerance horn) we should now specify. The connected line of plots for KE positions should not be wildly irregular, but rather deviate in a rather smooth, consistent fashion as with the example in Figures 15b and 15c.

Please note that I have never defined the allowable tolerances for the disparity of focal planes for different zones of a mirror in terms of fractions of wavelengths of light. Instead, I have defined them in far more unequivocal terms, terms that lead one to an intuitive understanding of what tolerances mean. Simple geometry, algebra, *and extensive practical verification* have unequivocally validated these procedures for me.

Before leaving our introduction to Foucault testing, it will be interesting to survey a mirror with the knife-edge whose figure is very irregular. Figures 16a and 16b show the same mirror (actual mirror from my extensive files) from two different vantage points along its OA for the KE. Note how radically different this mirror looks with the KE located in widely disparate positions along the OA.

I have kept this treatment of Foucault testing to the very barest essentials. A much more exhaustive treatment is possible; however, a well-explained introduction to the subject for beginners is what has been most wanted.

Appendix I

Figuring the Paraboloid Using the Method of the Full Sized Tool

With this article we will explain the elementals of figuring the paraboloidal curve for the concave mirror for the reflecting telescope using a full sized tool- that is to say, using a tool that is the same size as the mirror being figured. This is the time honored, traditional method of parabolizing mirrors of modest size and focal lengths. Larger and shorter focal length mirrors are best treated by other, newer methods. We will consider those methods in a separate treatise on the subject of figuring the paraboloid.

We will assume the worker has successfully made his or her polishing lap directly on the surface of the tool (usually the glass tool is used as the substrate, or surface on which to form the polishing tool).

In parabolizing, we desire to shorten the radii of curvature zone wise, successively, inwards towards the center of the mirror. That is to say, we will want to deepen the center of the mirror, relative to its outer regions. We will assume that the mirror at first is spherical, that is to say, that its radius of curvature is the same across any diameter of the mirror. In order to deepen the center of the mirror relative to its outer regions we will of course have to remove glass selectively, and in a graduated fashion with regard to the outer regions of the mirror. In other words, we will want to apply more polishing action on the inner regions of the surface of the mirror than on the outer regions. However, at the same time, we will want to blend, or graduate this amount of selective deepening of the center of the mirror so that there is no abrupt transition in the radius of curvature of the mirror, but rather, the radius of curvature of the mirror will gradually and smoothly increase when we measure zones successively outwards towards the edge of the mirror. Stated conversely, the radii of curvature will be gradually shorter, as we measure zones successively inwards towards the center of the mirror.

Method Explained

In order to graduate the amount of work applied across the face of the mirror so that we apply more work in the near central regions of it, we will vary the amount of surface of the mirror in contact with the polishing tool. In all cases except special instances where a defect may require using the tool on top of the mirror, we will work exclusively with the mirror on top of the tool. The tool will be secured to the top of the work stand, face up, and the mirror will be stroked, face down, on the surface of the polishing tool.

The primary type of strokes that will be employed will be straight strokes. A straight stroke with the mirror centered on the tool would be called a "diametral" stroke; however, we will not have the mirror centered on the tool except for a small portion of the figuring work. The figuring process will consist of regimens of stroking with the mirror overhanging the edge of the tool, so that the outside regions of the mirror receive less polishing action than the inside (near central) regions of the mirror.

We will establish some fundamental "rules" for polishing (figuring) with the mirror overhanging the edge of the tool. Let us list these rules in tabular form, here.

We will work the mirror in "stations", or "sectors", when working a single regimen of overhanging strokes; after finishing that sector of the mirror, we will move on and work another sector in identical fashion.

We will in every case while working a sector of the mirror as it overhangs the tool, also be moving the mirror laterally during stroking. That is to say, as we move the mirror forwards and backwards in straight strokes across the tool, we will also impart a lateral motion left to right as we make the forwards and backwards strokes across the tool.

We will never allow the center of the mirror, as we stroke it on the tool, to approach very closely the edge of the tool.

Let us now consider an example of a single regimen of strokes on a single sector of the mirror, with the mirror overhanging the tool. The worker will stand in place and begin with the mirror centered on the tool. We will require that a small black dot be marked on the center of the back of the mirror, to help us visualize the required stroke patterns as we work. As we proceed with a description of the required motions, we will use this illustration, Figure 17, to help the

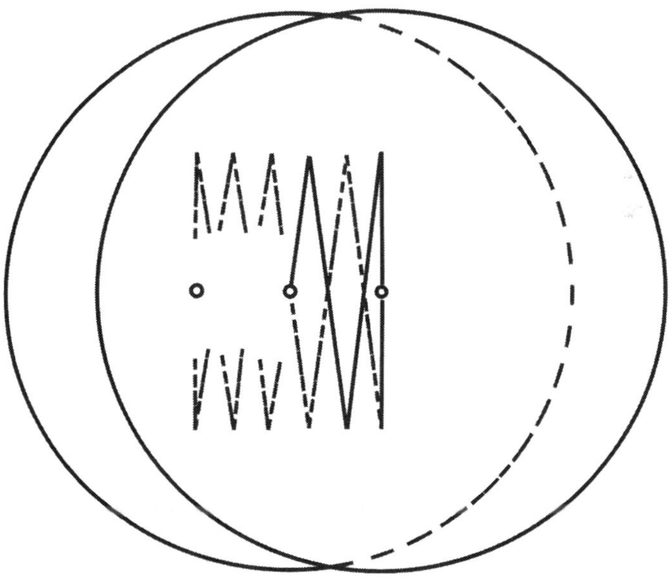

Figure 17.

worker visualize the required motions (and I will repeat this figure as necessary to reduce the need to return excessively to this page).

Now, looking again at Figure 17, the circle with a portion of its circumference represented as a dashed line represents the polishing tool underneath the mirror. The mirror is depicted in the figure as a circle with a solid line all the way around, and offset, or overhanging the tool by a small amount to the right. The pattern of "zigzags" represents, of course, a single regimen of parabolizing strokes for a single sector of the mirror. The three very minute circles you see lying across the raster of zigzagging strokes represent three different positions of the central dot on the backside of the mirror. It is by watching the movement of this dot that we will visualize our stroking patterns for each single regimen of strokes for any given sector we will apply a stroking regimen to.

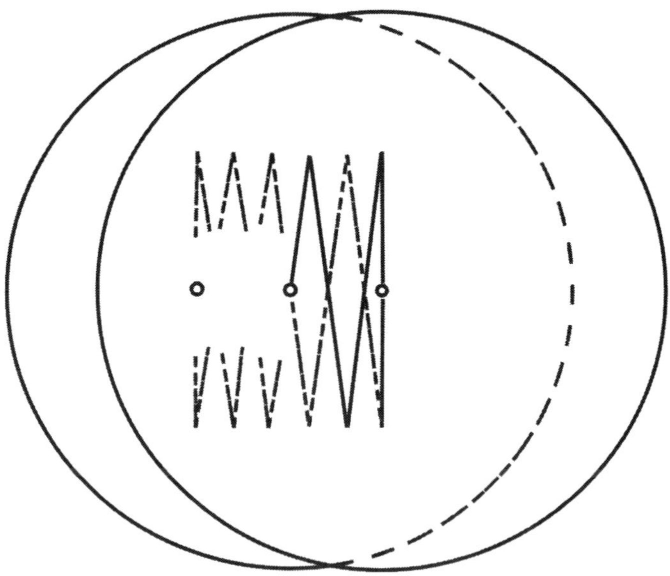

Figure 17 (repeated).

It is possible to very precisely control the developing paraboloid on the surface of the mirror without committing any errors by using this method. We will exercise perfect and complete control over the developing figure by varying the geometry of this approximately rectangular raster of zigzagging strokes. Most popular literature will refer to this kind of stroke as a "W" stroke; we need not specify the stroke verbally in this treatise, as you already see what kind of shape it has in the illustration. We will exercise perfect control by varying the parameters of three different aspects of the stroking regimen. I will list them in tabular form, here:

- We may vary the amount of lateral excursion (overhang) of the mirror during a single regimen of stroking a sector of the mirror.

- We may vary the speed, or rate of movement, of the mirror out to its limit of overhang during stroking. Most importantly, we may change the rate of lateral movement during the stroking regimen.

- We may vary the stroke length. This is not a very useful adjunct to controlling the developing figure, but we will have to vary the length of the stroke in any case for another reason, which will be explained shortly.

Now, we will proceed to describe in detail, using the same illustration, a single regimen of parabolizing strokes for a single sector of the mirror to be worked. Before we actually begin using the method of the full sized tool, we will want to also have in hand a method for keeping accurate track of which sectors of the mirror have been worked during a single round of the work stand (a single full round of work applied to all sectors of the mirror). The method of keeping track of work, sector wise, so that one does not mistakenly (accidentally) work a single sector twice, or alternatively, fail to work it even once during a single full round of working all sectors is called indexing. Indexing will be treated in a separate article. For now, we are primarily concerned with the method of actually forming the paraboloid.

The Forbidden Zone

Before actually commencing the parabolizing work, we will require to draw a circle on the back of the mirror in black ink. Center the circle on the central dot, Figure 18a; the circle, at first, should be approximately one third the diameter of the mirror. The interior of

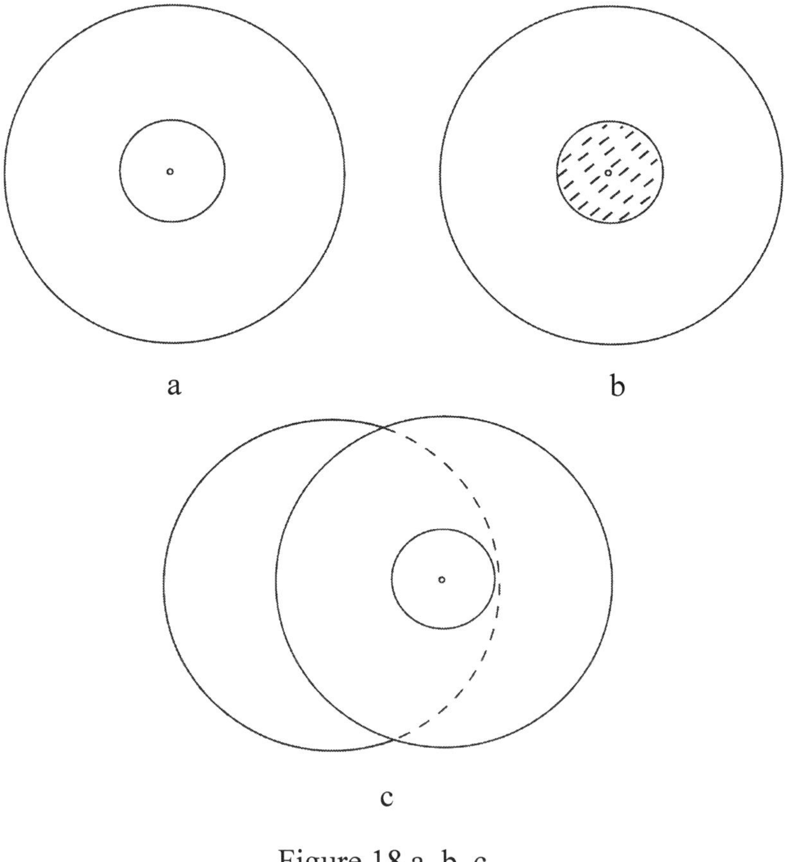

Figure 18 a, b, c.

this circle represents a forbidden zone, Figure 18b, for the edge of the polishing tool. That is to say, we will not allow the edge of the tool to be seen inside of this circle during any variation of stroking regimen, Figure 18c. The edge of the tool must not appear inside the forbidden zone.

Let us proceed to a description, now, of a single regimen of stroking for a single sector of the mirror. Look again at Figure 17 repeated below.

Now, after having the polishing lap well pressed and at a uniform ambient temperature (with the mirror also at a stable and identical ambient temperature) we place some abrasive slurry across the face of the polishing lap with a small brush dipped into the slurry, and place the mirror face down upon the tool, centered. In Figure 17, we

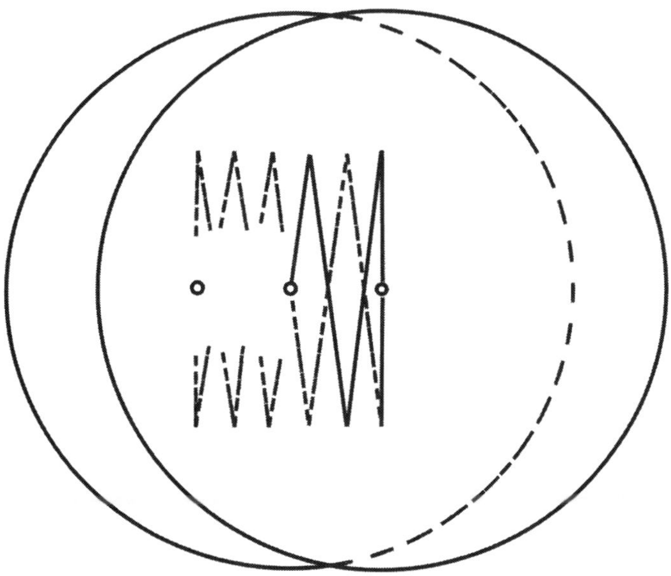

Figure 17 (repeated).

watch the central dot on the back of the mirror, as we begin our first stroke, the solid line starting from the minute circle (central dot) in the middle position representing the path of the line of the dot. We move the mirror away from, and slowly towards the right, to complete the first leg of the first stroke. Then, follow the path of the solid line as it descends downwards in the illustration, and continuing moving towards the right. After reaching its extreme downwards limit, the dot is then moved along the path of the solid line again upwards, and continuing its lateral excursion to the right, until we come again to the upper (furthest) limit of the stroke. From here, we begin a single downwards stroke, standing in place, with no lateral movement of the mirror for this single stroke. This last, single vertical stroke represents the extreme limit of excursion of the mirror laterally to its overhanging position on the polishing tool. In the illustrated case, the mirror is not overhanging very much at all; in fact, it is not overhanging enough to really effectively begin parabolizing. With this limited amount of overhang, the interior region of the mirror is not receiving very much more excavation (polishing) than the rest of the mirror. This is merely an illustration of the basic geometry of the stroking regimen. Let us continue as we describe the return path of the mirror from its limit of lateral excursion (overhang).

The return path stroke pattern is represented by dashed lines, and is identical to the pattern used when making the lateral excursion to the overhanging position on the tool. Follow the central dot, and note that the mirror may now stop at its starting point, having made (approximately) the same kind of zigzagging path returning, as it did on the way out to the side for the overhang.

You are now perhaps wondering about the other dashed lines, which seem interrupted in their middle portions, towards the left of the center of the illustration. These dashed lines indicate another alternative path for the stroking regimen. In our first description, we described a single regimen of stroking wherein we moved the mirror on an excursion from the center of the tool to an overhanging position, and then returned the mirror along an identical reversed path to stop it in the centered position again. However, instead of working a single sector of the mirror in this fashion, the left hand dashed lines show that, if we preferred to, we may continue the operation, identically, out to the left; that is to say, on our return from the right hand overhanging position, we may simply continue moving, not stopping with the mirror centered on the tool, but continuing and repeating the process on the opposite side of the mirror, thus, doing a right hand sector of the mirror, and continuing through the centered position towards the left hand side, and doing a left hand side of the mirror in a single operation, a single regimen of stroking, finally returning the mirror to the centered position on the tool.

Although this is the time honored tradition, the method most amateurs use (and also, they generally do not keep an accurate track of whether they have worked all sectors exactly once, without accidentally working a sector twice, or failing to work any single sector once), we will neither work opposite sectors in one operation, nor will we fail to index our work. Now let us consider altering our first example of a single regimen of stroking (i.e., the pattern of the regimen of strokes). Refer to Figure 19.

You will now observe that in the second example, the right hand example, we have depicted the mirror as having made a longer (further) excursion to its overhanging position. This will allow the center of the mirror to approach more closely the edge of the tool beneath it. Remember, however, that you have a circle drawn on the back of the mirror, which represents a boundary for which you must never see the edge of the tool inside of.

I must explain about keeping the stroke lengths the same. Many texts show the classic "W" strokes shortening as one approaches the extreme limit of the excursion of the mirror towards the edge of the

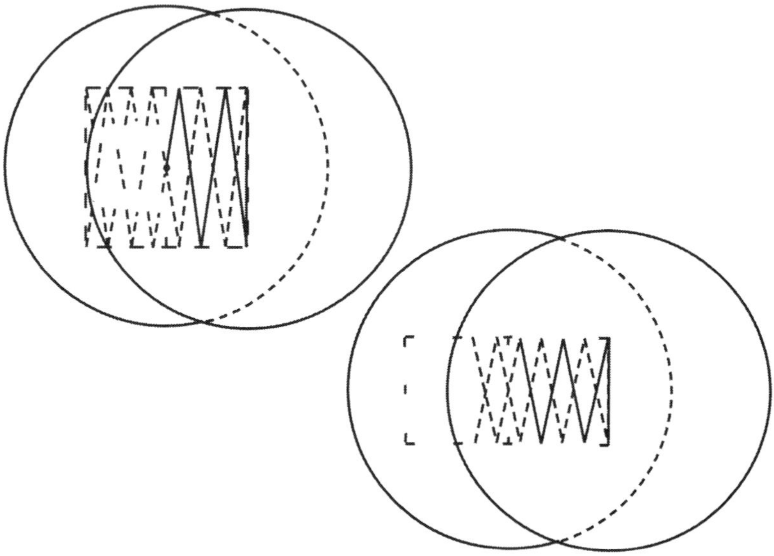

Figure 19.

tool, in order that the edge of the tool not end up dangerously close to the center of the mirror. Shortening of the strokes- that is to say, changing their length, is not allowed with the method I am teaching. Shortening the strokes as one approaches the edge of the mirror to keep its center a safe distance from the edge of the tool, can, and usually will, result in zoning. Therefore, we will adopt another method of avoiding allowing the center of the mirror to get too close to the edge of the tool when we are working a sector with a greater overhang, a greater lateral excursion of the mirror during stroking. What we will do is just keep all strokes necessarily short; it is not the length of the strokes that accomplishes the selective and graduated action of removing more glass from the central regions of the mirror, but the amount of overhanging outer regions receiving less glass removal through polishing.

Now we must discuss the two most important methods to employ in controlling the developing figure: the limits to the amount of overhang, and the variation in rate of movement during the lateral excursion of the mirror.

The limit of lateral excursion (the amount of overhang) will determine how aggressively we remove glass from the successively more central regions; the alteration of the velocity as we move the mirror outwards and then back inwards will give us precise control of the form of the developing figure.

Developing the Paraboloid: Control

The most satisfactory experience a beginning amateur mirror maker can have is to figure his or her mirror with such a technique or techniques that no errors of slope or defective zones (or any other kind of defect) develop at any time during the figuring process. The most common figuring error that the amateur will encounter is a relatively deep central depression, with the outer regions of the mirror remaining relatively spherical. The reason for this is that the worker has begun the parabolizing work too aggressively. Understanding this, it is logical to conclude that we must at least initially employ a very cautious and conservative approach when we begin the work of parabolization. In Figure 19, below or on page 74, we will note that the left hand example represents a less aggressive initial technique to advance the figure towards the paraboloid, because the extreme limit of lateral excursion of the mirror is not very much, relatively speaking, and also because all of the strokes, their lines as represented by the movement of the central dot tracing a path for us as we work, are equally separated from each other, laterally.

That is to say, each "leg" of each stroke, forwards or backwards, has the same "frequency"- the points forming the stop and reverse position for succeeding strokes, is approximately the same spacing relative to each other. I hope I have made this clear. Refer to the illustration again to get this clearly in mind- we present Figure 19 again, here:

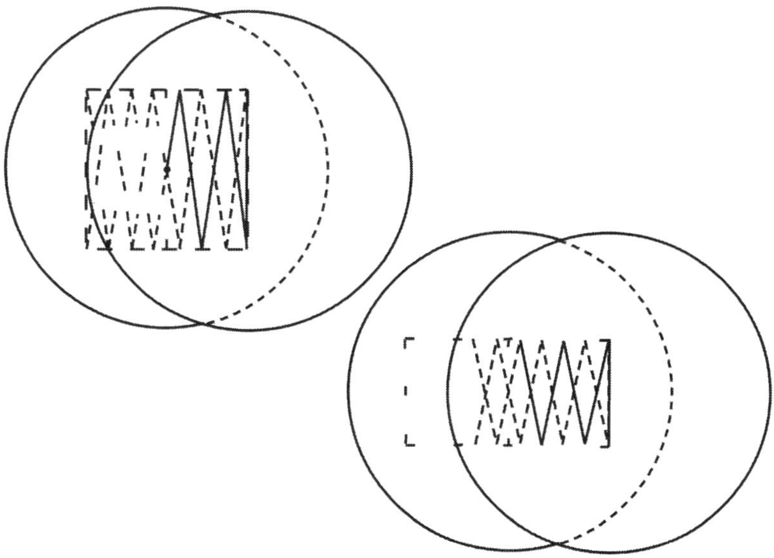

Figure 19 (repeated).

Now, if the mirror resists our efforts to deepen its center relative to its edge, the first remedy we shall apply is to reduce the size of the circle of our "forbidden zone" for the edge of the tool. Redraw this circle, to allow the center of the mirror to come closer to the edge of the tool. That is to say, to allow one's self to allow the lateral excursion (amount of overhang) of the mirror to the right to be greater. We will, as yet, not change the "frequency" of the strokes. Again, what we mean to say by this is that the distance between the strokes, laterally, will remain the same. We may keep these increments approximately the same by moving the mirror laterally at a relatively uniform rate of speed outwards toward the right as we work. When reaching the furthest desired limit of overhang for the mirror, one must always begin the return movement, laterally, at once. Working the mirror *in place*, without lateral movement as it is overhanging the tool is forbidden.

Avoiding the Central Depression

Somewhat paradoxically, as we begin parabolizing, we want to *avoid deepening* the center, but only in this respect: we want to avoid deepening it *too much* with respect to the outer zones. Instead we particularly want to endeavor to *blend* the deepened center continuously and smoothly with the successively outer zones of the mirror. That is why we will at first draw a circle on the back of the mirror representing the "forbidden zone" that we shall never allow the edge of the tool, seen through the back of the mirror, to pass into this forbidden zone.

I previously mentioned that this circle on the back of the mirror should be, initially, about *one third* the diameter of the mirror. It is likely that as you begin with "ordinary" or "regular" regimens of parabolization, that the mirror will seem to stubbornly resist any change in its figure. This is all right and is what we expect. It is part of the philosophy of approaching the work cautiously and safely. If we have made several rounds of the work stand, testing after one or two rounds of having worked each sector once, and the mirror has not shown any departure from its initially spherical figure, we will begin modifying our approach to more aggressively excavate the successively interior regions of the mirror.

Dealing with Further Resistance

Let us say that after having re-drawn the new, smaller circle on the back of the mirror to allow one a further lateral excursion (overhang) that the mirror is still resisting departing from spherical. The next step is that we shall abandon a uniform frequency of stroking. That is to say, as we work the mirror outwards towards its extreme limit of lateral excursion, we will *vary the speed* at which we move it outwards towards the overhang. If we keep up a regular "rhythm" of forwards and backwards stroking, but *accelerate* or *decelerate* the lateral movement of the mirror towards its limit of overhang, we shall in effect vary the "frequency" pattern of the strokes. Let me give two separate examples.

Let us say that we want to apply more strokes to the central regions of the mirror than to the outer regions, as in our hypothetical case of a mirror that is "resisting" parabolization. In that case, after having first tried the method of making the "forbidden zone" smaller; i.e., after having first tried a *greater* overhang of the mirror, we will next apply relatively more strokes to the central region of the mirror than we will to the outer regions to remove more glass to the central region, with this technique:

As we begin with the mirror centered, and begin our stroking regimen at the same time as we begin the mirror's lateral excursion out to the right, we will at first, as we begin the lateral movement, move it with relative rapidity. That is to say, we will begin our lateral movement *quickly*. As we begin to approach the edge, we will slow down the rate of lateral movement. If you visualize this, you will see that it results in *more* strokes being applied in the *central region* of the mirror during its overhanging increments. However, even though we have slowed down the rate of lateral movement so that the "legs" of the strokes are now closer together as the mirror completes its last small increments of lateral overhang (they may be characterized as having a higher "frequency"), we will under no circumstances disobey the previously stated rule: *we will never work the mirror unless it is moving laterally*, irrespective of the rate of change of that lateral movement.

I hope I have made it clear how "bunching up" additional strokes near the central region of the mirror will apply relatively more work there, while varying the lateral movement outwards, and then applying the same technique during the inward return from the lateral excursion (overhang) will allow one to apply more action to the central regions of the mirror, while at the same time blending the work smoothly.

Remember, when returning the mirror along its lateral path while stroking, that one will have to in effect do the opposite of what one did when moving outwards for the overhang- one must now begin returning the mirror, laterally, slowly, after having reached the furthermost lateral overhang, and then while proceeding progressively further inwards to center the mirror on the tool again, speed up the lateral movement.

Thinking about this, one can see that it is a very elegant technique for varying the amount of polishing action, zone wise, on the mirror.

Now, we have considered, in the above example, a case of where we have altered the rate of change of speed in the lateral movement so as to "bunch up" the strokes more closely together while the mirror's central regions are more nearly overhanging the tool. We may now consider an opposite case, where we intend to do exactly the opposite operation, and apply more work to the outer regions of the mirror relative to the central regions of the mirror.

Normally, one will not be required to ever use this "reversed" version of our parabolizing application of varying the rate of change of speed of lateral excursion (overhang) while figuring his or her mirror. One may expect, during the parabolization process, to gradually approach the correct curve in many increments of work, stopping after short working sessions and testing, to find, initially, how fast the work is progressing and therefore how long one may work before requiring testing again.

However, it is possible to use this technique of varying the acceleration of lateral movement in this opposite sense, contrary to our theoretical presentation of the case above where we wanted to apply even more aggressive polishing action to the near central regions of the mirror. However, oddly enough, using the technique in reverse will more often result in an irregular figure, than in directing the figure backwards again in the direction of sphericity. In case one has "overshot", and has worked the paraboloid too deeply, the best case is to return to regimens of polishing with the lateral excursion very short (very little overhang) and no acceleration or deceleration of the lateral excursion (frequency of strokes). Also, in the case of working remedial measures for a mirror whose paraboloid has become too deep (excessively eccentric) it is helpful to work the tool on top, with the same caveat: short strokes, very little lateral excursion (overhang of tool) and an even rate of lateral excursion (frequency of strokes). In summary, then, we have two powerful tools for parabolizing with efficient control. They are:

- Amount of lateral excursion of mirror (overhang), and
- Change of frequency of strokes (through changing rate of acceleration of lateral excursion, or overhang, of mirror).

In the event of a seriously overdone paraboloid, there are far more efficient and speedily accomplished remedies to reduce the degree of eccentricity of the figure; inasmuch as these fall under the realm of "area formatted" ("ring lap") polishing tools, we will not elucidate the methods in this treatise, but perhaps in another one.

The method of the "area formatted" lap is the most efficient method, allowing the most precise control, of any method for figuring the paraboloid; but again, that is matter for another treatment. Herein, above, we have covered the basics of figuring the paraboloid with the very old tradition of the full diameter tool, and it is very efficient and safe for the production of a fine paraboloidal figure for a mirror of long or moderate focal length. I have figured mirrors as short as f/4.5 with the method of the full diameter tool, but only small ones. The method is more and more problematical for mirrors of larger diameter with short focal ratios- these mirrors are more traditionally treated with sub diameter polishing tools for figuring, though they also may be treated with a full sized area formatted lap. The area formatted lap is relatively well known, but for some reason, not as popular as one might expect to be the case. It is my preferred method for figuring large and fast mirrors, though I confess that I can with equal facility and convenience use the method of the sub diameter polisher. Each of these latter two methods involves about an equal amount of work to use, and each allows very precise control, but the method of the sub diameter tool seems to have generally for the most part pre-empted the method of the area formatted polisher.

Appendix II

Figuring with the Area Formatted Ring Lap

There are three basic primary methods for figuring an aspherical mirror for astronomical use. The reason that no single one of these methods has supplanted either of the others is that each method is efficient for a particular range of size and focal length mirror. We may tabulate the three methods here, specifying the characteristics of the kind of mirror each method is the most efficient for.

(1) The method of the full sized (diameter) polishing tool. This method is particularly efficient, giving good control during figuring for small and medium sized mirrors, with long or moderately shorter focal lengths. The method may even be used for relatively "fast", smaller mirrors; I have figured a ten-inch diameter f/4.5 mirror with this traditional method without any untoward difficulties. For large and very large mirrors, it is a more problematical approach, even risky.

(2) The method of the small, or "sub-diameter" polisher. This method is particularly efficient, giving excellent control during figuring for large and very large mirrors, and is nearly ideally suited for large mirrors with very short focal ratios, f/5 and below. It is the ideal method of choice for amateurs working large glass nowadays.

(3) The method of the "area formatted" polishing tool. This method is particularly efficient for medium sized mirrors. Its most ideal application is in remedying zonal defects, although one may also use it to actually figure the paraboloid. We will explain in some detail the third method with this treatise, with particular emphasis on using the method for repairing zonal defects. To introduce the concept of the "area formatted" polishing tool, I will give an

illustration of an example of this kind of tool as used by George W. Ritchey for parabolizing.

Figure 20.

Figure 20 is from "On the Modern Reflecting Telescope, and the Making and Testing of Optical Mirrors", George W. Ritchey, Yerkes Observatory (No. 1459), City of Washington, Published by the Smithsonian Institution, 1904. This fine manuscript has been widely published and archived and is principally an account of the design and construction of the 60" reflector at Mount Wilson Observatory. The paper includes a detailed account of the work of producing the telescope's Cassegrain optics.

The tool represented by Figure 20 is a "full diameter tool". That is to say that the tool has the same diameter of the mirror that will be (was) worked with it. Ritchey did not use it exclusively in his work of parabolizing the 60" mirror; he gives account in his treatise of also using sub-diameter tools. I suspect that he found the full sized area formatted tool in Figure 20 rather tedious in use because its action can be anticipated to be slow. Perhaps he used it in the latter stages, as he "crept up" through the last stages of his approach to the correct figure for the 60" mirror.

The reason an area formatted tool like Professor Ritchey's is not popular now for figuring large, fast mirrors, is exactly the reason given in the paragraph above: the tool has a large amount of area in contact with the mirror, reducing the rate of removal of glass (we may observe that the larger the area of lap contact with any given mirror, the slower will be the removal of glass from any particular area of the mirror). And since the 60" mirror was both large and fast, a relatively large amount of glass was required to bring the mirror to its correct figure. So I suspect that Ritchey made his initial approach to the paraboloid aggressively with sub-diameter tools principally, and used this six lobed area formatted full sized tool to finish slowly, with fine control, and help smooth any zoning left over from his sub-diameter work.

The area-formatted lap is most ideally suited when one uses it to correct zonal defects. Usually one will have relatively little area of the polishing lap in contact with the mirror when using an area formatted lap this way, and consequently its action on the glass will be quite rapid. We may establish as a principal that with one of these kinds of laps, the more area in contact with the mirror's surface, the slower will be the rate of glass removal, and with lesser area in contact, the faster will be the rate of removal.

Let us consider the concept, now, in detail. The next illustration, Figure 21, will show an "apparent cross section" of the mirror, and

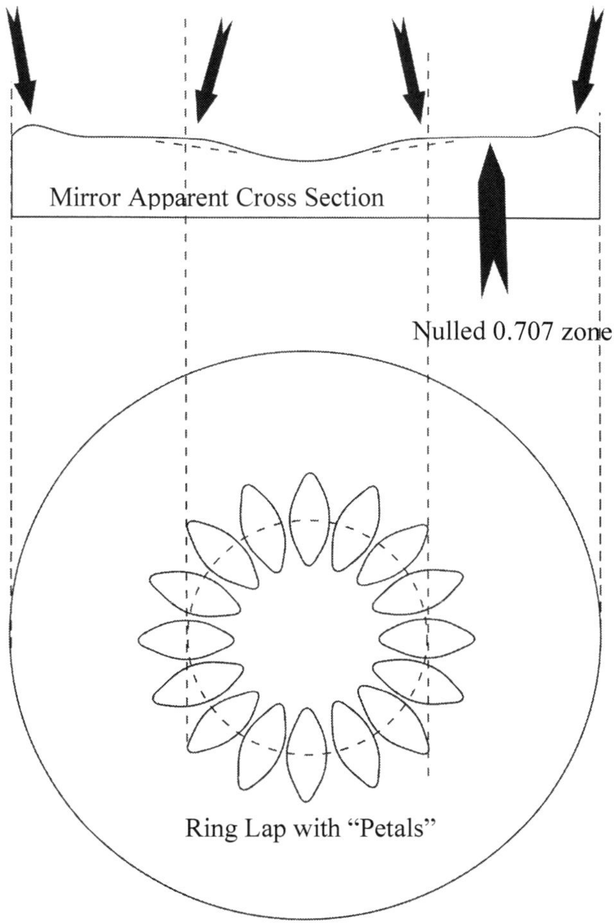

Figure 21.

below it and in line with it, a face on view of an area formatted polishing tool for remedying one of the mirror's zonal defects. This mirror may be considered spherical, but with two zonal defects. The first one, commonly obtained by the beginner, is a central depression from an overly aggressive, rather than cautious, approach to the paraboloid. In the apparent cross section, two short dashed lines below the surface of the mirror note this defect. For the moment we will ignore the other defect, the raised zone near the mirror's edge.

In the plan view of the polishing lap, one will note the arrangement of the circle of facets that will be in contact with the mirror; we will explain the very simple trick of temporarily transforming a full sized tool with all lap area in contact with the mirror into this area formatted tool, and then the simple method of returning the tool to a full sized polisher again after remedying the zonal defect.

Very often the shapes of the lobes in contact with the mirror will take on the shape you see both in Figure 20 and in Figure 21. They seem to me to be a shape similar to the African Zulu warrior's shield, and many times they will take a shape that resembles a spear point (blade). For this reason, I call this method of treating zonal defects "Zulu Figuring". Call it whatever you want to.

In this particular example, we will treat the central depression first, and then treat the anomalous zone near the mirror's edge. It is possible to treat both defects at the same time, with another ring of lobes on the same tool. Later, I will present an example of such an endeavor. After inspecting Figure 21 we will describe the allowed strokes for working the zonal defect.

Figure 21 clearly represents the principal of the selective application of polishing on the defective zone, and we need not elaborate here, further, finding Figure 21 sufficient to convey the concept. We will now proceed to explain the proper techniques in using an area formatted lap, following with some examples of mirrors representing a variety of zonal defects, and their corresponding area formatting ring laps for remedying those defects. After getting these concepts well in mind, we will explain the very quick and convenient method of modifying the full sized polisher into an area formatted lap, or what I prefer to call a "Zulu lap".

Using the Area Formatted Lap for Treating Zonal Defects

There will be occasions when the beginner may have difficulty discerning with certainty the location of the zone requiring treatment, or recognizing in an unequivocal manner how to go about treating it. For this reason it is handy to reposition the knife-edge to visualize all possible "apparent cross sections" that the repositioning of the knife-edge can show. Let us give an example by way of illustration of a mirror for which the knife-edge seems appropriately set, for getting a null for the largest region of the mirror. In this case it reveals a central depression in an apparently otherwise spherical mirror. I will first represent the mirror as it appears when the knife-edge is nulling the greater area of the mirror, instead of the central depression. Then we will show a different apparition of the mirror after having advanced the knife-edge toward from the mirror by a small increment, revealing a quite different apparent cross section. Let us then look at Figure 22a.

The central depression is glaringly evident in this focograph. However, we will now move the knife-edge towards the mirror to get the apparition as represented in Figure 22b:

Figure 22b, with the knife-edge in this position, we can now see clearly where to remove the glass to return the mirror to spherical. I drew this illustration conceptually, and recognize that in a real case such as this, the remnant central depression as depicted in Figure 22b. would probably not really be nearly so evident; that is to say, it would probably appear less severe, more shallow, than I have it represented here. But the concept remains the same. We now see that we have a broad, raised zone to polish down. We need to note where the crest of this zone is located, as a radius of the mirror, in order to know where to make our "Zulu spear blades" the widest. They will of course taper on both sides; and since the raised zone on this mirror is not symmetrical (it slopes more gently towards the perimeter of the mirror, and more steeply towards the center) we will draw out spear blade shaped lobes accordingly on our stencil for making a lap to treat this defect.

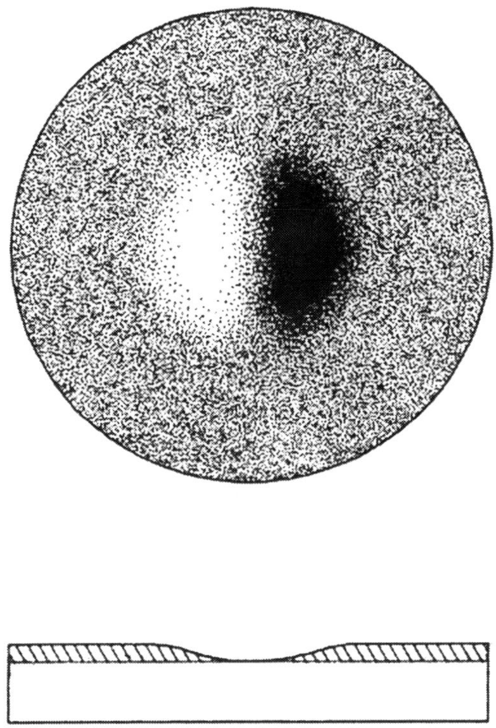

Figure 22a.

When laying out a stencil to form a Zulu lap, you should think of the shape of the lobes as representing the shape of the apparent cross section (cross hatched, in the drawing) but doubled, so that they will be symmetrical. It is not necessary to make the spear blades with exactly the same dimensions as the apparent cross section; but one must draw them with their proportions congruent with the zone to be

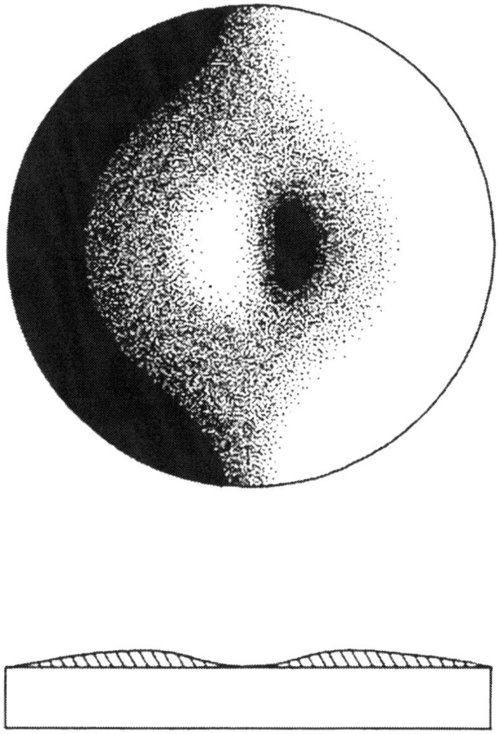

Figure 22b.

worked on the mirror. It is advantageous to make these spear blades exaggerated in their proportions, so that the action on the mirror will be a maximum action. After looking at a few more examples of stencils for treating zones, we will describe the making of an area formatted lap and also the proper technique for using one for remedying a zonal defect.

Let us now consider the proposition of a mirror with a turned down edge, the bane of many a beginner. The turned down edge used to be referred to as the "fatal turned down edge", because early in the

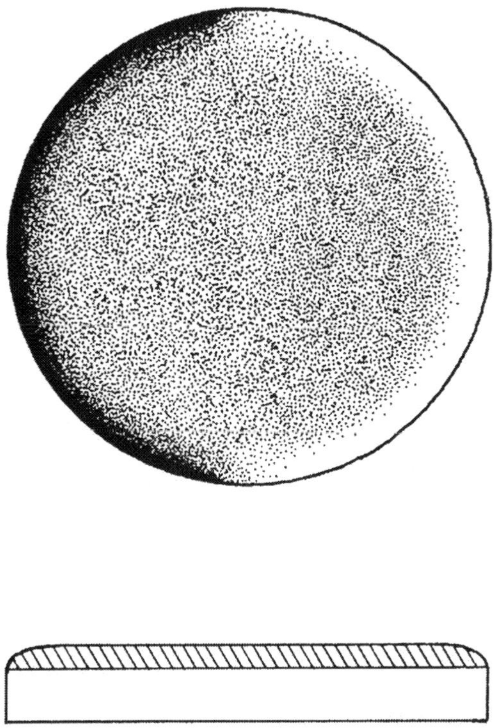

Figure 23a.

history of mirror making with glass, turned down edge seemed to be so commonly inevitable on polish out, and seemed to be so difficult to eliminate. Ideally one will finally develop his or her technique to prevent one from forming during polish out. Incredible as it may seem, I almost always get a turned up edge every time I polish out a mirror prior to figuring it. I have even approached my work adopting a technique designed to get a turned down edge, and incredibly, usually end up with a turned up edge anyway.

In Figure 23a we show a focograph and its apparent cross section, revealing what appears at first to suggest material we will need to remove to lower the entire surface down to the level of the edge of the mirror.

From Figure 23a, it would seem that if we reduced the surface down to the edge of the mirror (since the edge appears turned-down), we would then have a perfectly spherical mirror (the whole surface will appear flat when nulled). The difficulty with this approach is that it is hopelessly impossible to accomplish! Instead, we will revisualize the mirror in such a way as to reveal the turned down edge as a <u>raised zone</u>.

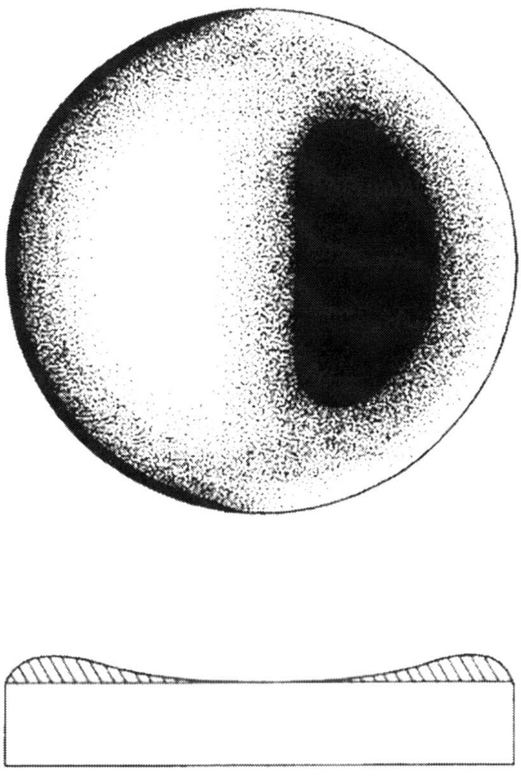

Figure 23b.

In Figure 23b we have moved the KE well beyond the Center of Curvature of the mirror in order to revisualize where the least amount of material needs to be removed to restore the mirror to its correct figure.

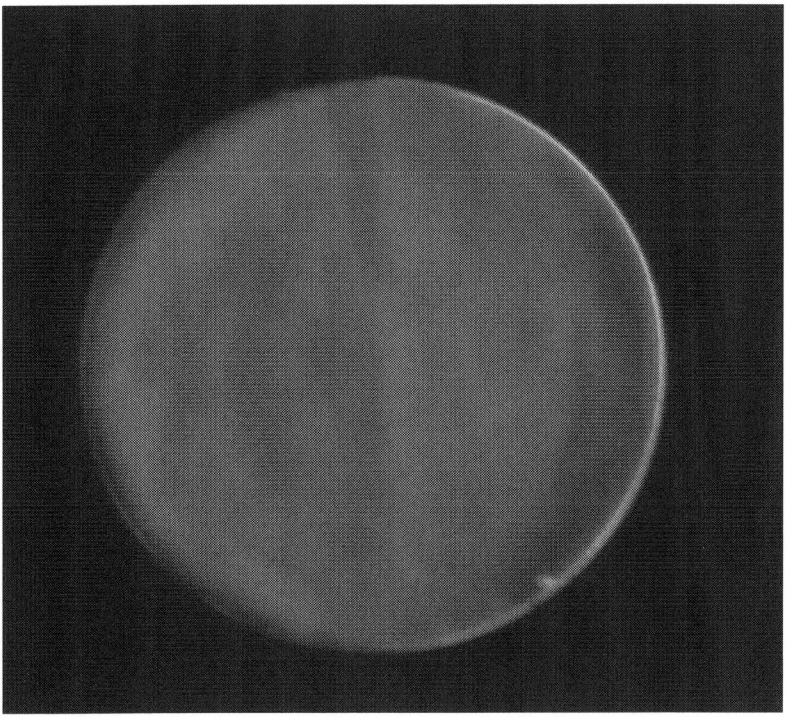

Figure 24.

A turned down edge may be a broad defect, wide, with a not excessively sharply turned down slope at the perimeter of the mirror, as in my illustration, above, but more often it is in fact the case that the turned down edge will be narrow and its outside slope will be very sharply turned down. Let us consider by example a mirror which does have a very sharply turned down edge; in this example, the mirror is generally spherical, but as the edge is approached, it begins to have an rather gently upward sloping zone that suddenly turns sharply downwards at the edge. In Figure 24, from Pat Bunn, we show a focograph (a true photographic focograph) of this real world example. Pat Bunn generously supplied this focograph, and I thank him for his contribution. The apparent unevenness of the surface of the mirror is due to air currents. In Figure 25, below, we show a stencil design for making a ring lap to treat this defect.

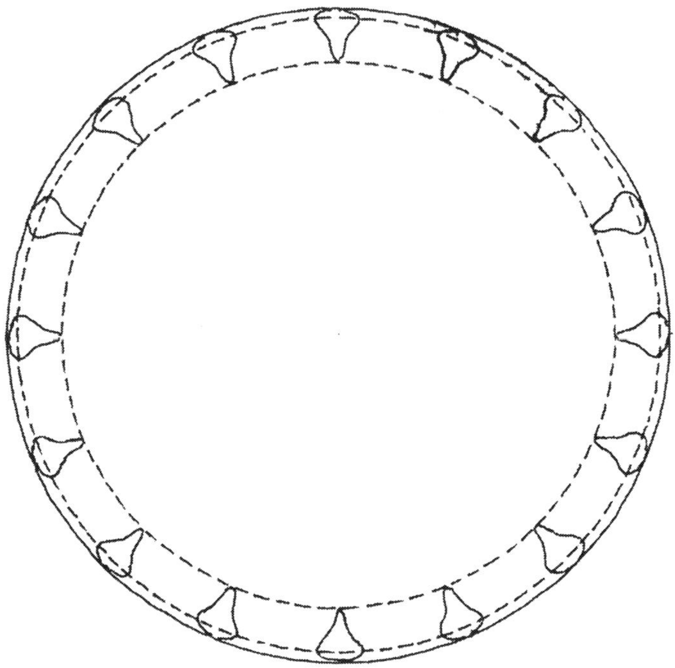

Figure 25.

One should carefully note that the lobes are configured in a fashion that reflects the defect on the surface of the mirror, in which the surface turns up gently before turning down abruptly. The stroke for treating this defect is generally a “forbidden stroke”, as it can lead to zoning, but we are already trying to treat a zone, so it is allowed: one may merely rotate the mirror (or tool) in place to plane down the defect. If finer sub-zones should develop, we will hopefully be able to reduce them with a following smoothing operation with the full sized lap.

Treating Multiple Zones in a Single Operation

Here is another focograph, Figure 26, also by Pat Bunn, revealing a mirror with defects.

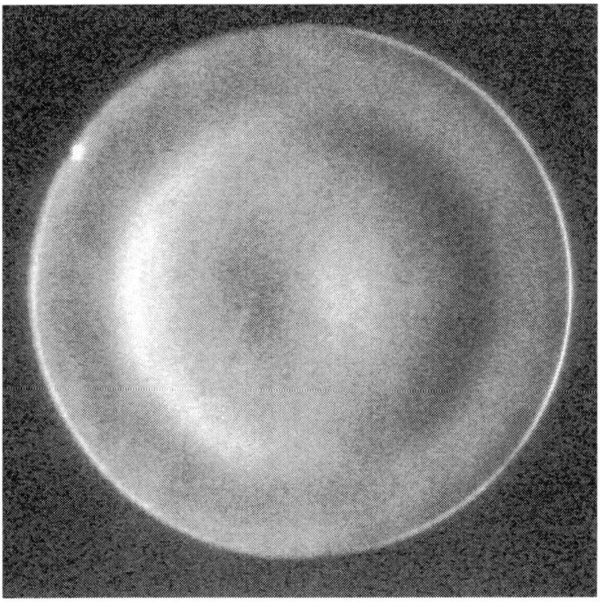

Figure 26.

On the next page, we show the stencil used to prepare a multiple ring lap to treat all of the defects that are glaringly evident in this focograph (raised central mound, and raised zone slightly over half way out; the edge is pretty fair, so we will not do anything to it yet).

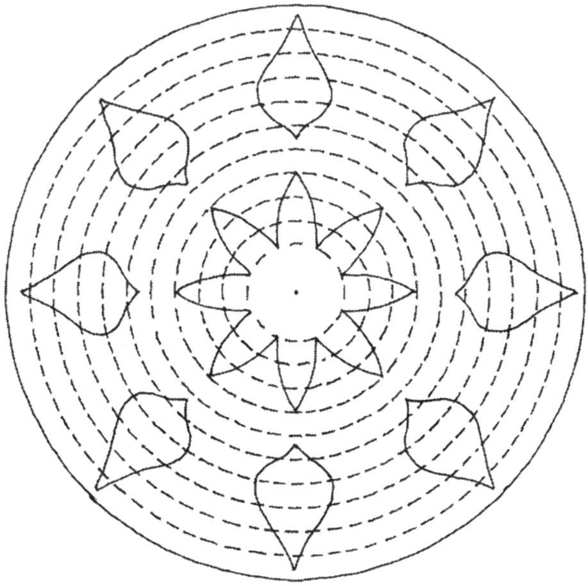

Figure 27a.

Figure 27a is the stencil we will use to prepare an area formatted ring lap to treat both of the defects simultaneously. The correct stroke for treating the defect is again the tiny circular stroke; watch the little dot on the center of the back of the mirror, and make it draw a very small circle, about five sixteenths of an inch in diameter. Naturally, observe the other required motions: change work stations around the barrel often, and every now and then rotate the mirror fractionally.

How to Make the Area Formatted Ring Lap

Making the area formatted ring lap is as easy as falling off a log, or eating cherry pie. What we want to do is have the raised pattern on the pitch of our ring lap, or what I have come to call the "Zulu Lap" because so often the lobes of our area formatted lap will resemble

Zulu warrior shields, or their spear point blades. I present here, a rather “Generic Example” of a typical Zulu, or area formatted ring lap, Figure 27b.

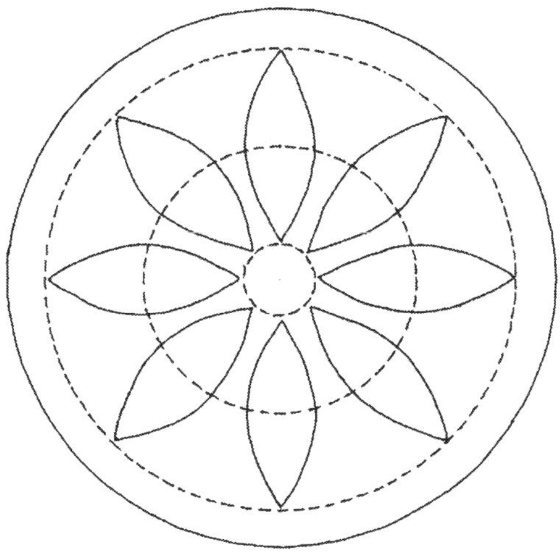

Figure 27b.

Method, Explained

You will have no trouble with this, as it is very much like making a pitch lap; in fact, it is easier; the lap is already made and all you have to do is modify it.

The neat thing about this is after you are done using your area formatted lap, you can restore it to its full area again (only having to re-channel, perhaps).

(1) Draw your appropriate stencil using a beam compass and pencil or pen, on the kind of thin poster board used in elementary school; you can find these thinner than Crescent board sheets everywhere; office supplies,

department stores, etc. Then, with a hobby knife number and 11 blade, cut out the lobes on your circular stencil made of this thin poster board.

(2) Fill the kitchen sink with water at about the temperature you used when you originally prepared your pitch laps; about 110 degrees F (or around 100 degrees).

(3) Immerse mirror and tool in this water; let the tool especially get thoroughly warmed up. Tool and mirror should be about same temperature.

(4) Pull mirror and tool (with its full diameter pitch lap on it still) out of the water, and put the stencil on top of the tool, centered, and place the mirror face down on the tool/stencil.

(5) Next, pick them up and squeeze them together with both hands; grip them along the edge, with thumb on top and fingers underneath, and squeeze the assembly ***hard, while they are still warm from the sink bath.***

Here is what will happen: the stencil will ***push down the polishing tool's entire lap everywhere except where your cut-out "Zulu Shields" are*** - these only will make contact with the mirror, because the cutouts allowed the pitch to squeeze through these cut out openings.

Now, after you have done this, if you want to get to work right away - then consider: you already have warm pressed. After taking the assembly apart, you should refill the sink with water at or near ambient temperature (temperature of the air in your shop). Your tool and mirror will cool off much quicker this way then if you put it back on the stand, and wait for air to cool it off; water draws the heat out quickly. ***Do not make the mistake of going right out and polishing;*** the temperature is still not yet homogenously distributed in the mirror. Put them on the work deck, both tool and mirror, on top, with no weight on the back of the mirror, and give them a good hour or so. It goes without saying that you should slurry up the tool first, and the face of the mirror and cover the assembly with plastic, as it becomes homogenous in temperature. ***Polishing with the mirror in temperature transition in any way will result in uneven polish***; usually resulting in a "lumpy" polish.

Allowing the temperature to reach ambient will ensure that your mirror and tool are truly ready for healthy polishing during your figuring sessions.

Note: Though the final appendix deals with "Indexing" during polishing, no indexing is required for using Zulu Laps . Polishing with a Zulu lap is for the express purpose of repairing a zone, and to do so only requires the mirror to be rotated over the tool.

Now, I would wish you good luck; though there is no such thing as luck, or accident, or coincidence; you are in control. Mirror making (figuring) is 99% conceptual. Use your head, and you are there- just think of cause and effect; that is the essence of figuring the paraboloid, there are no hard and fast rules set in concrete that you ever have to observe. Keep this motto in mind:

"Fortune Favors the Bold"

Final Note

These narratives on figuring the paraboloid are completely useless without a method to standardize the figuring strokes in such a manner as to apply the polishing action uniformly around and across the mirror surface, and in a manner that is completely repeatable while not impacting a zone of the mirror more, or less, than any other portion. All workers who desire to achieve as precise results as possible should *index* their work to achieve this uniformity and repeatability. Most probably do “index” their work, sector wise, in a more or less “slap-dash”, or not very precise or disciplined manner; and their mirrors turn out satisfactorily well, though with much rework required. However, the best opticians approach their work with iron discipline, and *index* their work precisely.

Therefore, the final appendix is a thorough narrative on the method of indexing. Using the disciplined method of indexing will enable you to distribute the stroking work over the mirror exactly equally, circumferentially, around the mirror. Using the method described in Appendix I, that is, the classic overhang method of parabolization with the full sized tool, it is important to ensure the mirror surface is worked uniformly a sector at a time. When grinding, one can easily lose track of how far the mirror has been rotated on the tool, and whether or not a sector of the mirror has been worked or not and to avoid working one sector too much or too little.

Indexing provides a completely controlled and reproducible approach to making sure that one does not work any one sector more than once, or not at all, in one complete round of work.

Appendix III
Indexing

We need an idiot-proof way to know how to distribute our figuring work equally around the mirror and tool, circumference-wise. So we will make a series of numbered index marks around the rim of our work stand's deck, and around the rim on the back side of our mirror.

Look at Figure 28. It shows a view of the top of a work stand with 12 index marks around its rim, and a mirror on the work deck with matching index marks all around its circumference. This scheme is set up for a <u>right-handed person</u>.

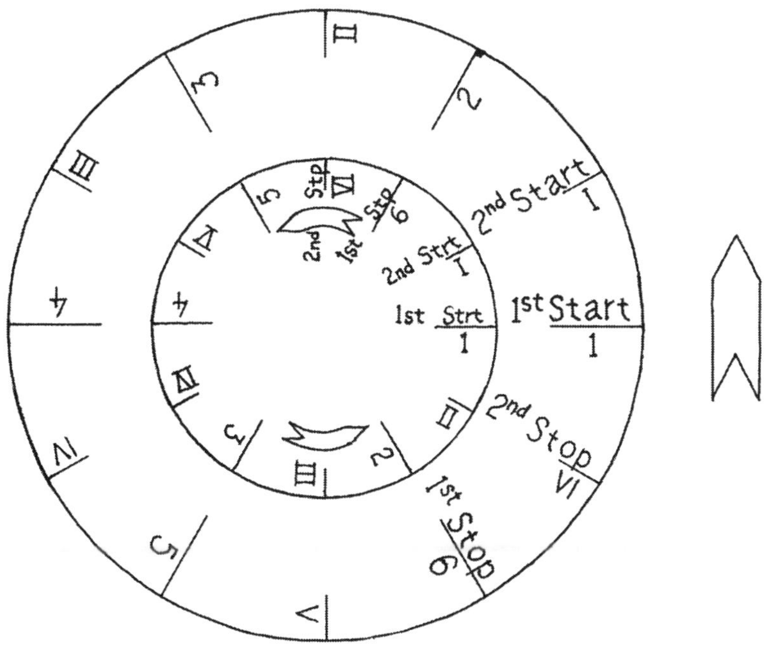

Figure 28. Outer circle: Index marks on the work stand.
Inner circle: Index marks <u>on the back side</u> of the mirror blank.

In Figure 28, the arrow to the right of the work deck, is to remind us of three things:

1) All side-throw excursions of the mirror to overhang the tool will be to the right only
2) We will always move counter-clockwise around the work stand
3) We will always be facing the direction indicated by the arrow when we do any single figuring increment at each indexed location

With this system we will be able to know always where started, where we stopped, with respect to both tool and mirror. We will not put index marks on the tool as we will expect that we will always be working with the mirror on top during figuring. In other words, the tool, being affixed to the work desk, will always retain its orientation matching the index marks on the work deck.

Note that the work deck is numbered (i.e. the index marks are numbered) counter-clockwise. The *long* index marks are numbered with Arabic numerals (1, 2, 3, 4,…); the alternate *short* index marks are numbered with Roman numerals (I, II, III, IV,..). Refer again to Figure 28.

A "full round" of figuring will be defined as one complete circuit of the work stand. The process of figuring is accomplished in ***six*** work increments: stopping and working in place, for the first full round of figuring, at each Arabic numbered index location. You will ***stand in place*** and work the mirror out to the right *while* stroking forward and backward, to overhang the tool by the desired (specified) amount of side-throw, and then return it in an exact reverse of motions back to centered on the tool again. (This stroking motion is described more clearly in the chapter on Figuring the Paraboloid).

Then, you will move around the work stand to the next Arabic numbered index mark, turn the mirror also counter-clockwise until its matching (same) Arabic numbered index mark is lined up adjacent to this next index mark on the work stand.

To clarify, let me refer to the next set of illustrations. As stated above, the drawing is of the work deck, with the mirror centered on the tool. Where you stand is seen in Figure 28a, below.

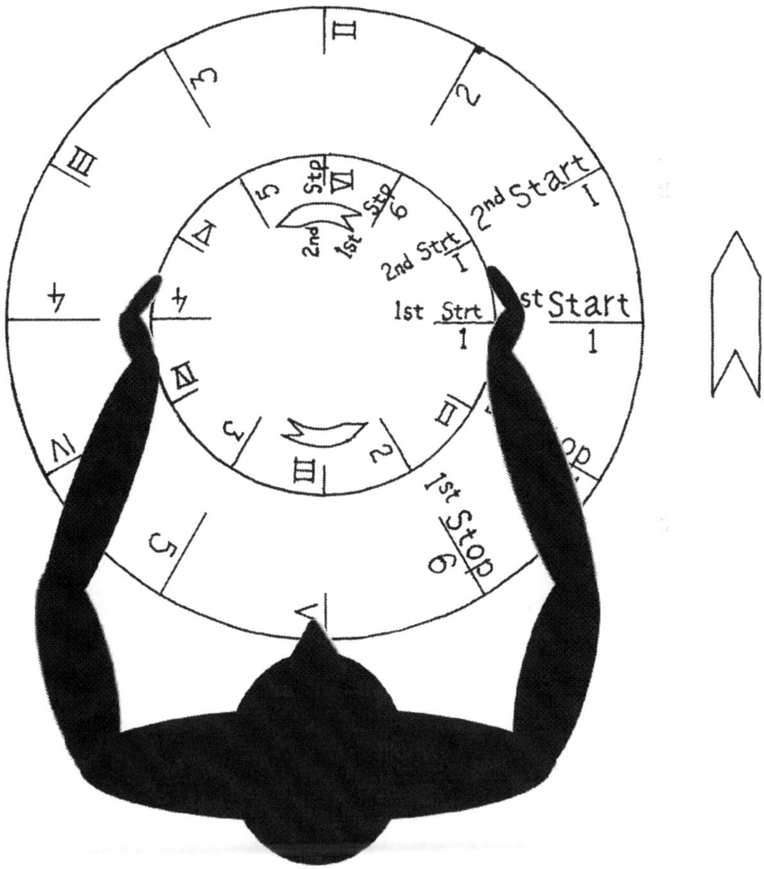

Figure 28a.

The illustrations are always presented with the observer looking down from above. You are looking down on the top of the work stand with the mirror's "**1st Start**" lined up with the work deck's "**1st Start**" position. You are facing the arrow to the right of the drawing, and your right hand is on the edge of the mirror where "1st Strt" is labeled. Your left hand, of course, is on the left side of the mirror, over the upside down "4". The big arrow illustrated at the right side of the stand always indicates the direction you are facing.

Now, refer to Figure 28b. It shows how you are to work the mirror. The work deck is not shown, but you are still facing the same direction and have the mirror and work deck lined up as in Figure 28a. Work the mirror exactly as shown – forward and back, out to your right, and then back to the center again. The motion creates the zig-zag pattern illustrated by the *solid* lines between the central dots.

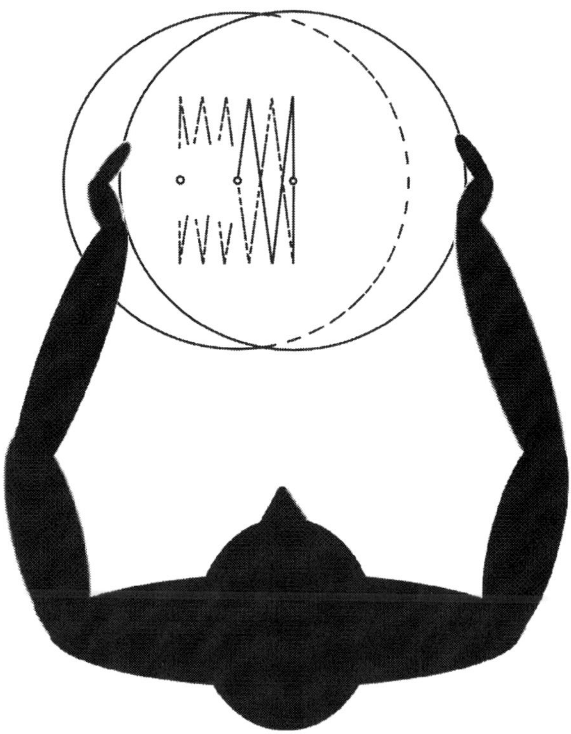

Figure 28b.

(The dashed lines in the zig-zag pattern of Figure 28b are described on pages 72-73).

In the next paragraphs, upon completion of the strokes in this position, you step to the right, counter-clockwise from the perspective of the illustrations, and turn the mirror, so the next regimen of strokes is done starting with the mirror's #2 Start

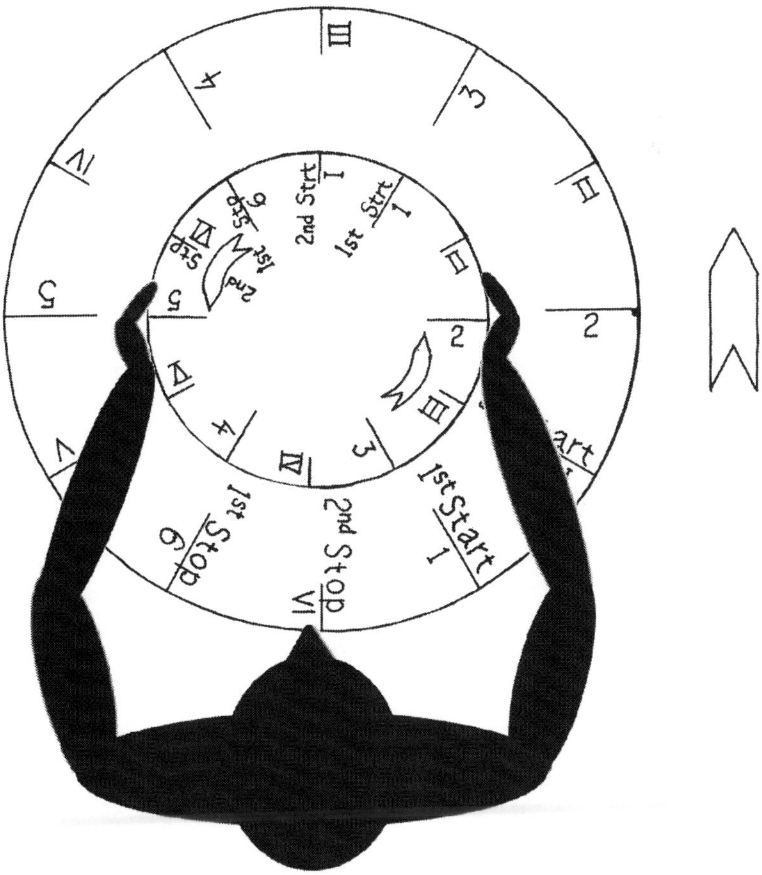

Figure 28c.

position, lined up with the work deck #2 position. This is seen in Figure 28c. Thus the narrative which follows, which describes a "full round" of stroking should become clear.

Observe the difference in Figure 28a and Figure 28c. In the latter, you have stepped to the right of the work deck, and your right hand is now lined up with the work deck's #2. From here, the stroking pattern illustrated in Figure 28b is repeated.

Process Illustrated

Now, to reiterate, we are looking down on the work deck with its mirror in place on the top of the tool, and we have placed the mirror so that its number one index mark, also labeled "1st Start" (abbreviated "1st Strt" in Figure 28 or 29a) is adjacent the work stand's number one index mark, also labeled "1st Start".

We will make our first increment of one <u>full round</u> of figuring right here, making our forwards and backwards strokes of the mirror on the tool, as we bring the mirror <u>out to the right</u>, towards the #1 ("1st Start") index mark as we continue stroking forwards and backwards, and then, when reaching the specified amount of overhang on the tool, we will without delay, begin to reverse this excursion of the mirror back again as we continue stroking forwards and backwards until we have the mirror centered again and at rest on the tool, just as before we made this first one sixth (1/6) increment of one "full round" of figuring. Then, we walk to the <u>next</u> <u>Arabic</u> numbered <u>index location</u>, #2, on the rim of the work deck, and turn the mirror counter-clockwise until it's <u>#2 index</u> mark is <u>adjacent</u> the <u># 2 index mark</u> on the work deck.

Referring to Figure 29 (or Figure 28c, previous page), we see how we've relocated our position counter-clockwise around the work deck to position 2, and turned (rotated) the mirror also counter-clockwise, to its position 2.

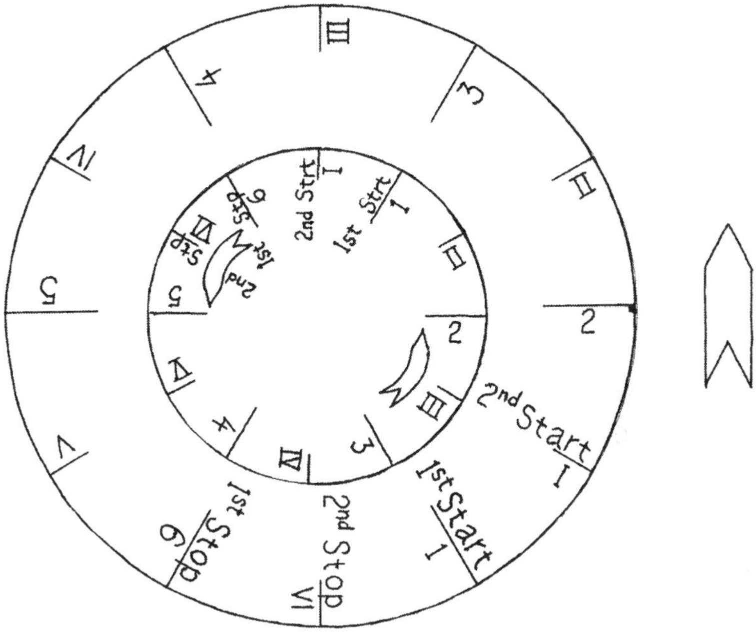

Figure 29.

Now, standing in this position, we will accomplish our 2nd increment of one full round of figuring, by duplicating our steps made when we figured while standing in place at the #1, or "1st Start" position, making our figuring strokes and bringing the mirror <u>out to the right</u> for the specified amount of overhang, and then returning it again as we continue stroking to finish this increment with the mirror again centered on the tool, with both #2 index marks lined up.

Now, I think you can guess what comes next! We now move our position to the work deck's #3, and rotate the mirror to it's #3 (Figure 30), and repeat the pattern of strokes (Figure 28b, page 104).

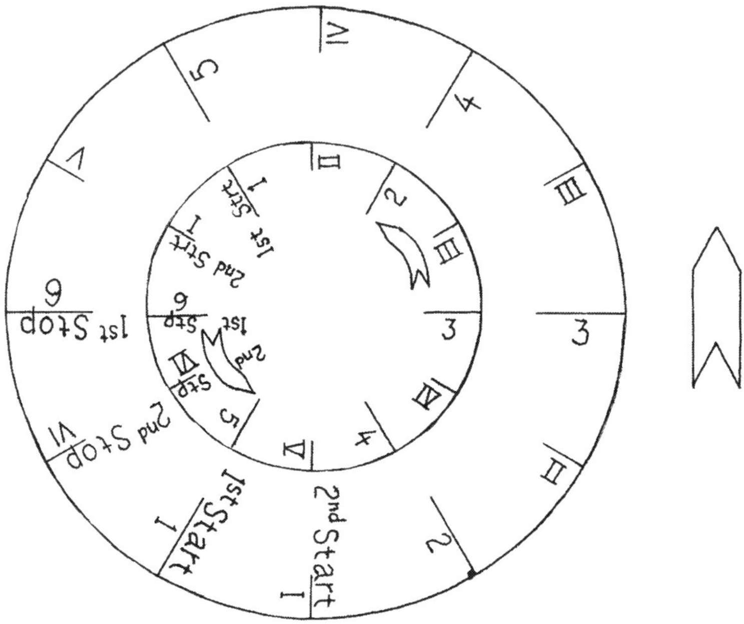

Figure 30.

In this manner, you will progress to sector #4, #5, and #6. You will notice that when we have finished our last incremental round of figuring, the #6 position, #6 is labeled “1st Stop”, Figure 31. This indicates we have correctly distributed ***six*** sectors of figuring *equally* around the mirror, and have not overlapped (or underlapped) the work distributed around the mirror.

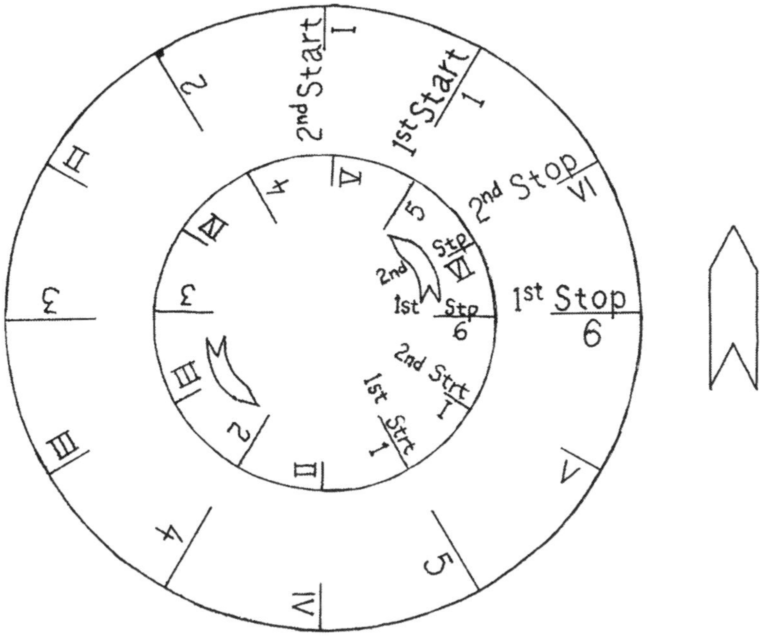

Figure 31.

For the next full round of figuring, we will repeat all previous steps, starting with everything lined up at the "2nd Start" position, which you will notice, are begun on the sectors marked with the Roman numerals. In this way, the sectors of the mirror have been divided into not just 6 sectors, but in 12, distributing our strokes homogeneously over the entire mirror surface, significantly improving the development of our parabola much more uniformly.

Another reason we have two different sets of sectors is so the mirror gets at least 12 sectors worked in such a way that the abrasive is not depleted as much as if you worked all twelve sectors with one charge of slurry. In other words, this system will give the mirror more even work, as described above, not only to the mirror surface, but to the distribution of the slurry. Note that instead of Roman numerals, you could mark these alternate sectors with letters. The importance is that you clearly mark *two different* set of sectors.

Remember never to work, that is "stroke", with the mirror standing in-place as I have admonished repeatedly. You *must* be *moving* the mirror to the right (or left), as you stroke, creating the described zig-zag. Also, *do not let the edge of the tool be seen through the back of the mirror inside the circle that defines the forbidden zone*. (See page 70 for a description of the forbidden zone.)

At first your mirror will give you resistance in parabolizing; you have started with a one-third diameter circle "forbidden zone". If tests reveal little change, and indicate that you are not digging a hole, make the circle smaller, say, one-fourth the diameter of the mirror. If you still get resistance (that is, tests show the sphere is not deepening as desired), change your strokes from evenly spaced to closer spaced as the mirror's *central* dot approaches the edge of the tool (slow down your lateral movement) and then when heading back in to return the mirror on top of the tool, start back in slowly, as you slowed down on the way out, to keep the strokes closer together while the mirror's center is near the edge of the tool. Remember, *under no circumstances are you allowed to stop the lateral motion of the mirror and stroke in-place*. The motion must always be forward and back (away from you and toward you), while moving the mirror laterally either to the right ("off" the tool), or toward the center (to re-center the mirror on top of the tool). On the way back toward the center, speed up to space the strokes a little wider apart, as they were when you started out a little faster, before slowing down.

Appendix IV
Anticipating the Outcome

I tried to think of a way to provide more samples to illustrate all the aspects of parabolizing a mirror. There are so many variables, and one will typically not produce more than 2 or 3 mirrors in their lifetime, so honing the skills needed to examine the zonal images and to understand what the image says is hard to come by! Using a nice software produced by Michael Lindner called, appropriately enough, "Foucault Simulator V0.1" we can now apply what we have learned for some comparisons.

Thinking now of the 6 essential foundation lessons, and how the KE is used to examine the 5 "zones", let's compare the focographs of 3 mirrors, each 12" diameter, but with 3 different focal lengths (and hence 3 different radii of curvature.)

Here are the locations "d" (formula, Figure 10, page 48), for each of the 5 zones addressed in pages 46-48:

Mirror Diameter: 12"

	Focal Length 60"	Focal Length 72"	Focal Length 84"
Zone 0:	0.000	0.000	0.000
Zone 1:	0.015	0.012	0.011
Zone 2:	0.045	0.038	0.032
Zone 3:	0.075	0.062	0.054
Zone 4:	0.105	0.088	0.075
Zone 5:	0.134	0.112	0.096

In the table above, I have added a "Zone 0", which is, of course, by definition, the KE at the "design" radius of curvature, exactly twice the focal length, 120", 144", and 168". Thus the numbered zones are measured, in inches, from this "zero" starting point.

We can already make some interesting observations. First, since the 60" focal length mirror possesses a "deeper" curve to its surface, we might expect the distance between Zone 0 and Zone 5 to be closer, or "tighter" than less-deep mirrors of longer focal length. But close inspection of the position of Zone 5 for the 60" FL mirror shows it is 0.134 inches from Zone 0, and Zone 5 for the 84" FL mirror is only 0.096" from its Zone 0. How is this apparent discrepancy to be reconciled?

It is true that the "curve" of the 60" is "deeper" than the 84" – however, the 84", having a FL (and radius of curvature) 40% longer than the 60" FL mirror, creates a "shallower" convergence of the rays from the edge of the mirror, and hence, the distance "d" from Zone 0 to Zone 5 for the 84" FL mirror is reduced by (can you guess?) …. 40%. (Test this is you like. Increase the number for Zone 5 for the 84" FL mirror by 40%. What do you get?)

A second interesting observation is in the numbers representing "d" for Zone 1, for all three mirrors. Now these numbers are rounded to the nearest thousandth, but are remarkably close to each other. This is precisely why a dial screw or micrometer is recommended as the instrument of choice to accurately determine "d" in you Foucault test device. The total range over which the KE must translate is no more than ½ inch in most applications, but the accuracy should be readable in *at least* to 0.001 inch, preferably 0.0005 inch.

Now let us compare the results of the Foucault Test against these 3 mirrors. In each case, we will stop the KE at the respective zones for each mirror, and compare the results for each focal length. We do not want to compare the results at the same "d", for we d not use "d" to interpret the shadows, rather, we are always interpreting the null at each *Zone*.

I think that you will understand after examination the next sequence of illustrations, why the skilled mirror maker says: "Deepen the center of the sphere to create the paraboloid."

12" Dia, 60" FL 12" Dia, 72" FL 12" Dia, 84" FL

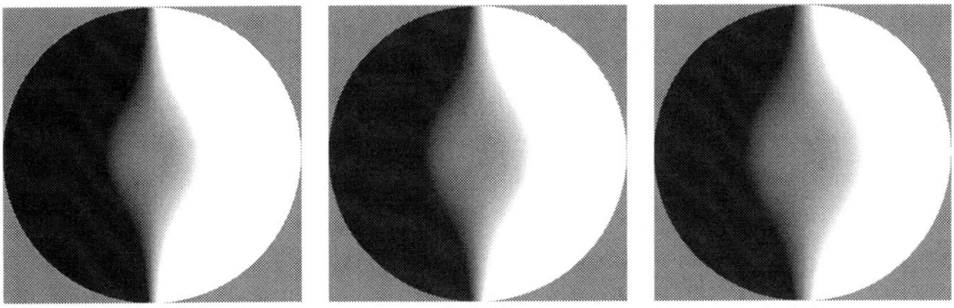

Zone 0: 0.000 0.000 0.000

Note that for the 60", the deeper center reveals a bit more "hole".

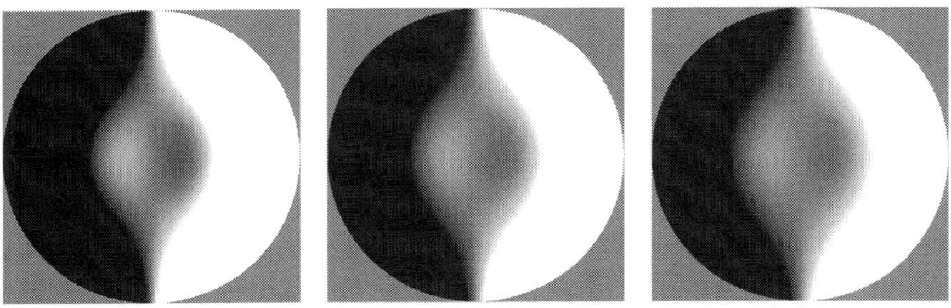

Zone 1: 0.015 0.012 0.011

Here, the 60", at Zone 1, already shows evidence of the "rolling crest" described on page 45.

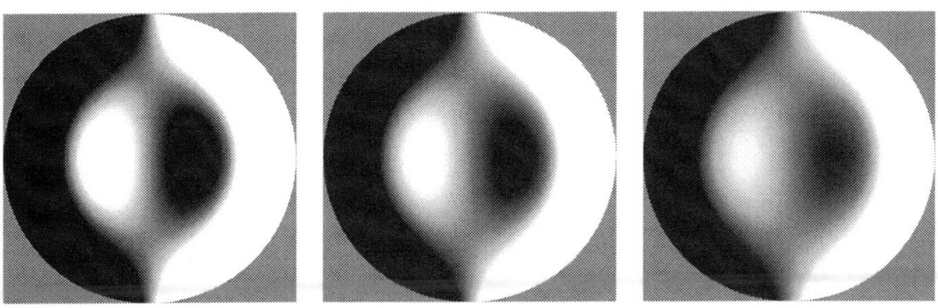

Zone 2: 0.045 0.038 0.032

Here, at Zone 2, the 84", though broader, shows the first evidence of the "donut".

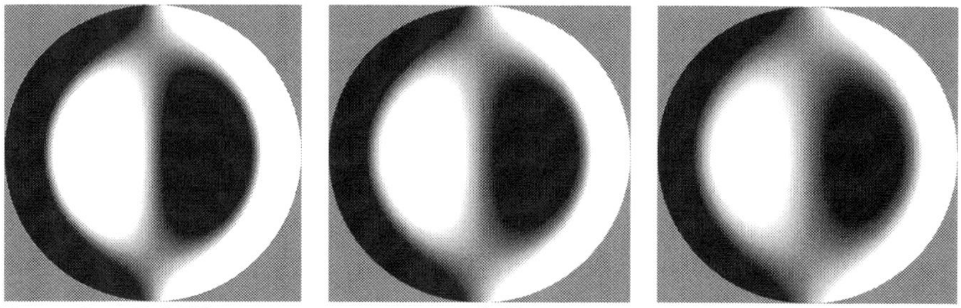

Zone 3: 0.075 0.062 0.054

Take note of the similarity now at Zone 3. You begin to see the essence of the paraboloid – how it shallows out toward the edge.

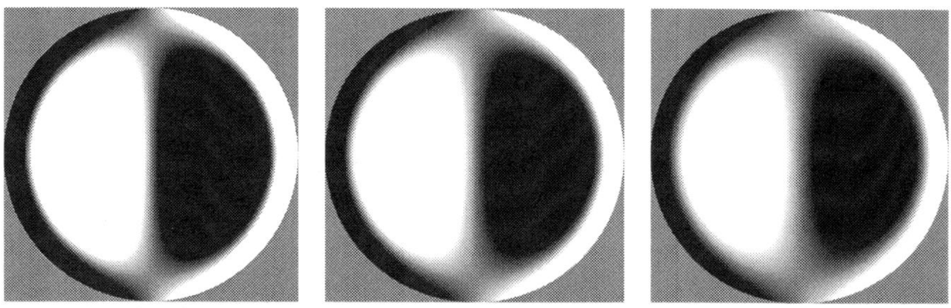

Zone 4: 0.105 0.088 0.075

Here, at Zone 4 you are hard pressed to tell the difference, yet the difference in focal length between the left and right images is 40%!

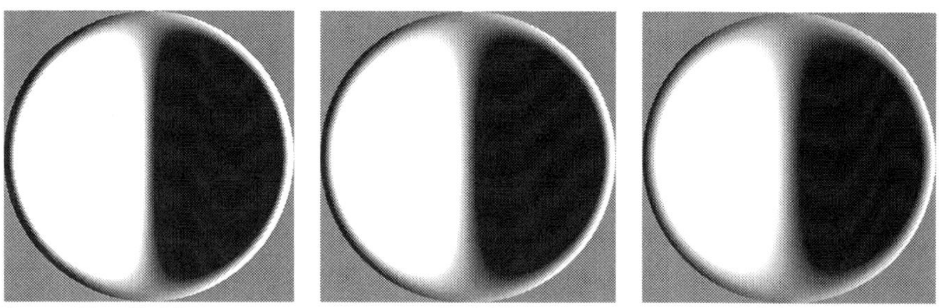

Zone 5: 0.134 0.112 0.096

And finally. Zone 5. Practically indistinguishable. The point, however, is that the evidence is the same – the Foucault Test reveals even out at Zone 5, how the paraboloid should appear.